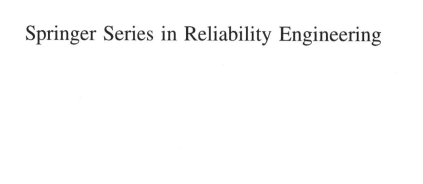

Springer Series in Reliability Engineering

For further volumes:
http://www.springer.com/series/6917

Dana Kelly · Curtis Smith

Bayesian Inference for Probabilistic Risk Assessment

A Practitioner's Guidebook

 Springer

Dana Kelly
Idaho National Laboratory (INL)
PO Box 1625
Idaho Falls, ID 83415-3850
USA
e-mail: Dana.Kelly@inl.gov

Curtis Smith
Idaho National Laboratory (INL)
PO Box 1625
Idaho Falls, ID 83415-3850
USA
e-mail: Curtis.Smith@inl.gov

ISSN 1614-7839
ISBN 978-1-84996-186-8   e-ISBN 978-1-84996-187-5
DOI 10.1007/978-1-84996-187-5
Springer London Dordrecht Heidelberg New York

British Library Cataloguing in Publication Data
A Catalogue record for this book is available from the British Library

Cover design: eStudio Calamar, Berlin/Figueres

Printed on acid-free paper

Springer is part of Springer Science+Business Media (www.springer.com)

Preface

This book began about 23 years ago, when one of the authors encountered a formula in a PRA procedure for estimating a probability of failure on demand, p. The formula was not the obvious ratio of number of failures to number of demands; instead it looked like this:

$$\tilde{p} = \frac{x + 0.5}{n + 1}$$

Upon consulting more senior colleagues, he was told that this formula was the result of "performing a Bayesian update of a noninformative prior." Due to the author's ignorance of Bayesian inference, this statement was itself quite noninformative.

And so began what has become a career-long interest in all things Bayesian. Both authors have indulged in much self-study over the years, along with a few graduate courses in Bayesian statistics, where they could be found. Along the way, we have developed training courses in Bayesian parameter estimation for the U.S. Nuclear Regulatory Commission and the National Aeronautics and Space Administration, and we continue to teach the descendants of these courses today. We have also developed and presented workshops in Bayesian inference for aging models, and written a number of journal articles and conference papers on the subject of Bayesian inference, all from the perspective of practicing risk analysts.

After having developed a Bayesian inference guidebook for NASA, a guidebook on Bayesian inference for time-dependent problems for the European Commission, and an update of a paper on Bayesian parameter estimation written in the 1990s, we were approached by Springer with the idea of writing a textbook. There is an ever-increasing number of Bayesian inference texts on the market, due in large part to the growth in computing power and the accompanying popularization of the Markov Chain Monte Carlo (MCMC) techniques we employ herein. Many, if not most of these texts are written at the level of an advanced undergraduate or beginning statistics graduate student. In other words, the available references are generally beyond the level of the typical risk analyst in the field, who is most often an engineer, and who may have had at most a course or two in probability and statistics along the way, typically from a frequentist perspective.

Having struggled without success to find a suitable text for our courses over the years, we wanted to write a text that would be accessible to a majority of practicing risk analysts. We wanted to employ the modern technique of MCMC, which can handle a wide range of what were once intractable problems. Having followed the development of BUGS since its inception in the 1990s, we decided to write the text around this software. There are other choices of software for this type of analysis, many of which are also free and open-source. JAGS (Just Another Gibbs Sampler) is one whose syntax is very similar to that of BUGS. The R software package has a number of MCMC routines available, as well as packages that interface with BUGS and JAGS, allowing them to be run in "batch mode." R also has packages for processing the output of BUGS and JAGS, including convergence diagnostics and graphics. The Python language has PyMC, which is the focus of much advanced development efforts these days. We encourage the interested reader to explore these software packages, along with others that will come along in the future.

BUGS has evolved into OpenBUGS, and most of the examples in this text were solved using OpenBUGS 3.1.2. At the time this text went to press, Ver. 3.2.1 was released. This version has significant enhancements not covered in this text, most notably ReliaBUGS, which includes a variety of specialized distributions used in advanced reliability analysis. It's probably time to begin work on a second edition!

We hope to remove some of the mystery that seems to surround formulas like the one above, and to make plain often-heard incantations such as "Bayesian update of a noninformative prior." For after all, Bayesian inference is, we feel, much more straightforward than its frequentist alternative, especially in the interpretation of its results. There is one formula, Bayes' Theorem, which underlies all that is done, and understanding the component parts of this theorem is the key to just about everything else, including the specialized jargon that has accumulated in Bayesian inference, as in every other field in science and engineering.

We begin the text in Chapters 1 and 2 with the motivation for using Bayesian inference in risk assessment, and provide a general overview before moving into the most commonly encountered risk assessment problems in Chapter 3, those involving a single parameter in an aleatory model.

Chapter 4 introduces the too-often overlooked subject of model checking, and illustrates a number of checks, both qualitative graphical ones and quantitative checks based on the posterior predictive distribution.

Chapter 5 introduces more complicated aleatory models in which there is a monotonic trend in the parameters of the commonly used binomial and Poisson distributions.

Chapter 6 discusses MCMC convergence from a practical point of view. In principle, convergence to the joint posterior distribution can be problematic; however, in risk assessment convergence is rarely an issue except in the population variability models discussed in Chapter 7. However, if there is more than one parameter involved, the prudent analyst will always check for convergence.

Chapter 8 introduces more complex models for random durations, covering inference for the Weibull, lognormal, and gamma distributions as aleatory models. It also introduces penalized likelihood criteria that can be used to select among candidate models, focusing on the deviance information criterion (DIC) that is calculated by OpenBUGS.

Chapters 8 and 9 together describe time-dependent aleatory models that are encountered when modeling the infant mortality and especially the aging portions of the famous "bathtub curve." Chapter 9 covers inference for the so-called renewal process, where failed components are replaced with new ones, or repairs restore the component to a good-as-new state. Chapter 8 covers the situation where repair is same-as-old; rather than reincarnating a failed component, the failed component is merely resuscitated. Chapter 9 also discusses some useful graphical checks for exploratory data analysis.

Chapter 10 turns to analysis of cases where the observed data are uncertain in some way, perhaps because of censoring, inaccurate record-keeping, or other reasons. The Bayesian framework can handle these kinds of uncertainties in a very straightforward extension of the case without such uncertainty.

Chapter 11 introduces regression models, in which additional information in the form of observable quantities such as temperature can enhance the simpler aleatory models used in earlier chapters.

Chapter 12 describes inference at multiple levels of a system fault tree. For example, we might have information on the overall system performance, but we might also have subsystem and component-level information.

Chapter 13 closes the text with a selection of problems, which are generally of a more specialized or advanced nature. These include an introduction to inference for extreme value processes, such as might be employed to model an external flooding hazard, an introduction to treatment of expert opinion in the Bayesian framework, specification of a prior distribution in OpenBUGS that is not one of the built-in choices, and Bayesian inference for the parameters of a Markov model, the last illustrating the ability to numerically solve systems of ordinary differential equations within OpenBUGS, while simultaneously performing Bayesian inference for the parameter values in these equations.

We apologize in advance for the errors that will inevitably be found, and hope that they will not stand in the way of learning.

Dana Kelly
Curtis Smith

Contents

Chapter 1
Introduction and Motivation

1.1 Introduction

The focus for applications in this book is on those related to probabilistic risk analysis (PRA). As discussed in Siu and Kelly [1], PRA is an analysis of the frequency and consequences of accidents in an engineered system. This type of analysis relies on probabilistic (i.e., predictive) models and associated data. Because of PRA's focus on low-frequency scenarios, often involving the failure of highly reliable equipment, empirical data are often lacking. Bayesian inference techniques are useful in such situations because, unlike frequentist statistical methods, Bayesian techniques are able to incorporate non-empirical information. Furthermore, from a practical perspective, Bayesian techniques, which represent uncertainty with probability distributions, provide a ready framework for the propagation of uncertainties through the risk models, via Monte Carlo sampling.

There are even more advantages in adopting a Bayesian inference framework in PRA. For example, in data collection and evaluation, we strive for the situation where the observed data are known with certainty and completeness. Unfortunately, for a variety of reasons, reality is messier in a number of ways with respect to observed data. For example, one may not always be able to ascertain the exact number of failures of a system or component that have occurred, perhaps because of imprecision in the failure criterion, record keeping, or interpretation. Further, when there are multiple failure modes to track (e.g., fails during standby versus fails during a demand), one may not be certain for which failure mode to count a specific instance of a failure.

Even though an estimate of a component's failure rate or failure probability is required for use in the PRA, a detailed analysis and data gathering effort is not always possible for every part/assembly. Consequently, PRA must use the available data and information as efficiently as possible while providing representative uncertainty characterizations—it is this drive to provide probability distributions

D. Kelly and C. Smith, *Bayesian Inference for Probabilistic Risk Assessment*,
Springer Series in Reliability Engineering, DOI: 10.1007/978-1-84996-187-5_1,
© Springer-Verlag London Limited 2011

representing what is known about elements in the risk analysis that leads us to Bayesian inference.

The hardware modeling element of the PRA has typically relied on operational information and data, coupled with Bayesian inference techniques, to quantify performance. This analysis of performance uses the Bayesian approach in order to escape some of the problems associated with frequentist estimates, including:

- If data are sparse, frequentist estimators, such as "maximum likelihood estimates," can be unrealistic (e.g., zero);
- Propagating frequentist interval estimates, such as confidence intervals, through the PRA model is difficult;
- Frequentist methods are sometimes of an *ad hoc* nature;
- Frequentist methods cannot incorporate "non-data" information into the quantification process, other than in an *ad hoc* manner.

The last limitation of frequentist inference listed above highlights a key attribute of Bayesian methods, namely the ability to incorporate qualitative information (i.e., evidence) into the parameter estimates. Unlike frequentist inference, which focuses solely on "data," the Bayesian approach to inference can bring to bear all of what is known about a process, *including* empirical data.

1.2 Background for Bayesian Inference

The Bayesian (or Bayes–Laplace) method of probabilistic induction has existed since the late 1700s [2, 3]. Laplace, starting in 1772, performed the first quantitative Bayesian inference calculations. The application then was inferring the mass of planets such as Jupiter and Saturn using astronomical observations, along with simple (i.e., uniform) prior distributions, but a rather complicated stochastic model [4]. Unfortunately, the Bayesian mathematics for Laplace's problem was quite complicated, primarily due to his selection of a particular stochastic model (a double exponential, also known today as a Laplace distribution).

Later, in 1809, Gauss popularized the normal (or Gaussian) distribution. Laplace, having been made aware of this new stochastic model, returned in 1812 to his previously intractable problem of inference about the mass of Saturn. To perform his Bayesian calculation, he used:

- Data (orbital information on the satellite Callisto).
- A prior (the uniform distribution).
- A less complicated stochastic model (the normal distribution).

What Laplace calculated was a posterior probability distribution for the mass of Saturn. He published his results in the *Théorie Analytique des Probabilités*, representing the first successful quantitative application of Bayes' Theorem.

Today, the Bayesian approach to inference is employed in a very wide variety of domains for many different stochastic modeling situations. For example, two widely used stochastic models (both in and outside of PRA) are the Poisson and binomial, representing different processes:

- Examples of Poisson processes.

  - Counting particles, such as neutrons, in a second.
  - Number of (lit) lights failing over a month.
  - Arrival of customers into a store on a Monday.
  - Large earthquakes in a region over a year.
  - HTTP requests to a server during a day.

- Examples of Bernoulli processes.

  - Tossing a coin to see if it comes up heads.
  - Starting a car to see if it will start.
  - Turning on a light to see if will turn on.
  - Launching a rocket to see if it will reach orbit.

The basis of many traditional PRAs is event tree and fault tree models (deterministic models), which logically relate the occurrence of low-level events to a higher-level event (e.g., an initiating event followed by multiple safety system failure events may lead to an undesired outcome). The occurrence of initiating events and system failures (or just "events") in the fault trees and event trees are modeled probabilistically, and the associated probabilistic models each contain one or more parameters, whose values are known only with uncertainty. The application of Bayesian methods to estimate these parameters, with associated uncertainty, uses all available information, leading to informed decisions based upon the applicable information at hand.

Most PRAs require different types of "failure models" to quantify the risk portion of the analysis. These models include the following:

- Failure of a component to change state on demand.
- Failure in time of an operating component.
- Rate of aging for a passive component.
- Failure (in standby) of an active component while in a quiescent period.
- Downtime or unavailability due to testing.
- Restoration of a component following a failure.

We describe Bayesian inference for these and other models in this book, using modern computational tools.

1.3 An Overview of the Bayesian Inference Process

In Chap. 2, we will introduce Bayes Theorem, which according to the theory of subjective probability, is the only way in which an analyst whose probabilities obey the axioms of probability theory can update his or her state of knowledge [5]. The general procedure for performing Bayesian inference is:

1. Specify an *aleatory model*[1] for the process being represented in the PRA (e.g., failure of component to change state on demand).
2. Specify a *prior distribution* for parameter(s) in this model, quantifying epistemic uncertainty, that is, quantifying a state of knowledge about the possible parameter values.
3. Observe *data* from or related to the process being represented.
4. Update the prior to obtain the *posterior distribution* for the parameter(s) of interest.
5. Check validity of the aleatory model, data, and prior.

We follow this process to make inferences, that is, to estimate the probability that a model, parameter, or hypothesis is reasonable, conditional on all available evidence. As part of describing the process of Bayesian inference, we used several terms such as "data," "aleatory," and "epistemic." In this text, we attach specific meanings to key terms for which confusion often exists:

Data   Distinct observed values of a physical process. Data may be factual or not. For example they may be subject to uncertainties, such as imprecision in measurement, truncation, and interpretation errors. Examples of data include the **number** of failures during part testing, the **times** at which a tornado has occurred within a particular area, and the **time** it takes to repair a failed component. In these examples, the observed item is bolded to emphasize that data are observable. An aleatory model is used in PRA to model the process that gives rise to data

Information The result of evaluating, processing, or organizing data and information in a way that adds to knowledge. Note that information is not necessarily observable; only the *subset* of information referred to as data is observable. Examples of information include a calculated estimate of failure probability, an expert's estimate of the frequency of tornados occurring within a particular area, and the distribution of a repair rate used in an aleatory model for the restoration time of a failed component

Knowledge What is known from gathered information

Inference The process of obtaining a conclusion based on what one knows.

To evaluate data in an inference process, we must have a "model of the world" (or simply "model") that allows us to translate observable events into information [6, 7]. Within this framework, there are two fundamental types of model

---

[1] Also referred to synonymously as a "stochastic model," "probabilistic model," or "likelihood function."

abstractions, aleatory and deterministic. The term "aleatory" refers to the stochastic nature of the outcome of a process. For example, flipping a coin, testing a part, predicting tornadoes, rolling a die, etc., are typically (chosen to be) modeled as aleatory processes. In the case of flipping a coin, the observable stochastic data are the outcomes of the coin flips (heads or tails).

Since probabilities are not observable quantities, we do not have a model of the world directly for probabilities. Instead, we rely on aleatory models (e.g., a Bernoulli[2] model in the case of tossing a coin) to infer probabilities for observable outcomes (e.g., two heads out of three tosses of the coin).

Aleatory   Pertaining to stochastic (non-deterministic) events, the outcome of which is described using probability. From the Latin *alea* (a game of chance or a die)

Deterministic   Pertaining to exactly predictable (or precise) processes, the outcome of which is known with certainty if the inputs are known with certainty. As the antithesis of aleatory, this is the type of model most familiar to scientists and engineers and includes relationships such as $V = IR, E = mc^2, F = ma, F = G\, m_1\, m_2/r^2$

In PRA, we employ both aleatory and deterministic models. In these models, even ones that are deterministic physical models, such as thermal–hydraulic models used to derive system success criteria, many of the parameters are themselves imprecisely known, and therefore are treated as uncertain variables. To describe this second type of uncertainty (with aleatory uncertainty being the first kind), PRA employs the concept of epistemic uncertainty.

Epistemic   Pertaining to the degree of knowledge about models and their parameters. From the Greek *episteme* (knowledge)

Whether we use an aleatory model (e.g., Bernoulli process) or a deterministic model (e.g., $F = ma$), if any parameter in the model is imprecisely known, then there is epistemic uncertainty associated with the output of that model. It is the goal of this book to demonstrate how to combine data and information with applicable models and, via Bayesian inference, enhance our knowledge for PRA applications.

---

[2] A Bernoulli *trial* is an experiment whose outcomes can be assigned to one of two possible states (e.g., success/failure, heads/tails, yes/no), and mapped to two values, such as 0 and 1. A Bernoulli *process* is obtained by repeating the same Bernoulli trial, where each trial is independent of the others. If the outcome given for the value "1" has probability p, it can be shown that the summation of n Bernoulli trials is binomially distributed $\sim$ binomial (p, n).

References

1. Siu NO, Kelly DL (1998) Bayesian parameter estimation in probabilistic risk assessment. Reliab Eng Syst Saf 62:89–116
2. Bayes T (1763) An essay towards solving a problem in the doctrine of chances. Philos Trans R Soc 53:370–418
3. Laplace PS (1814) A philosophical essay on probabilities. Dover Publications, New York (reprint 1996)
4. Stigler SM (1986) The history of statistics: the measurement of uncertainty before 1900. Belknap Press, Cambridge
5. de Finetti B (1974) Theory of probability. Wiley, New York
6. Winkler RL (1972) An introduction to Bayesian inference and decision. Holt, Rinehart, and Winston, New York
7. Apostolakis GE (1994) A commentary on model uncertainty. In: Mosleh A, Siu N, Smidts C, Lui C (eds) Proceedings of workshop I in advanced topics in risks and reliability analysis, model uncertainty: its characterization and quantification. NUREG/CP-0138, U.S. Nuclear Regulatory Commission, Washington

Chapter 2
Introduction to Bayesian Inference

2.1 Introduction

As discussed in Chap. 1, Bayesian statistical inference relies upon Bayes' Theorem to make coherent inferences about the plausibility of a hypothesis.

Observable data is included in the inference process. In addition, other information about the hypothesis is included in the inference. Consequently, in the Bayesian inference approach, probability quantifies a state of knowledge and represents the plausibility of an event, where "plausibility" implies apparent validity. In other words, Bayesian inference uses probability distributions to encode information , where the encoding metric is a probability (on an absolute scale from 0 to 1).

Note that the use of the word "hypothesis" here should not be confused with classical Neyman-Pearson hypothesis testing. Instead, the types of hypotheses that might be evaluated when performing PRA include:

- The ability of a human to carry out an action when following a written procedure.
- The chance for multiple redundant components to fail simultaneously.
- The frequency of damaging earthquakes to occur at a particular location.
- The chance that a component fails to start properly when demanded.

2.2 Bayes' Theorem

Bayes' Theorem provides the mathematical means of combining information and data, in the context of a probabilistic model, in order to update a prior state of knowledge. This theorem modifies a prior probability, yielding a posterior probability, via the expression:

D. Kelly and C. Smith, *Bayesian Inference for Probabilistic Risk Assessment*, Springer Series in Reliability Engineering, DOI: 10.1007/978-1-84996-187-5_2, © Springer-Verlag London Limited 2011

Table 2.1 Components of *Bayes'* Theorem

Term	Description
$P(H|D)$	Posterior distribution, which is conditional upon the data D that is known related to the hypothesis H
$P(H)$	Prior distribution, from knowledge of the hypothesis H that is independent of data D
$P(D|H)$	Likelihood, or aleatory model, representing the process or mechanism that provides data D
$P(D)$	Marginal distribution, which serves as a normalization constant

$$P(H|D) = P(H)\frac{P(D|H)}{P(D)}. \qquad (2.1)$$

If we dissect Eq. 2.1, we will see there are four parts (Table 2.1) :

In the context of PRA, where we use probability distributions to represent a state of knowledge regarding parameter values in the PRA models, Bayes' Theorem gives the posterior distribution for the parameter (or multiple parameters) of interest, in terms of the prior distribution, failure model, and the observed data, which in the general continuous form is written as:

$$\pi_1(\theta|x) = \frac{f(x|\theta)\pi_0(\theta)}{\int f(x|\theta)\pi_0(\theta)d\theta} = \frac{f(x|\theta)\pi_0(\theta)}{f(x)}. \qquad (2.2)$$

In this equation, $\pi_1(\theta|x)$ is the posterior distribution for the parameter of interest, denoted as θ (note that θ can be a vector). The posterior distribution is the basis for all inferential statements about θ, and will also form the basis for model validation approaches to be discussed later. The observed data enters via the aleatory model, $f(x|\theta)$, and $\pi_0(\theta)$ is the prior distribution of θ.

Note that the denominator in Eq. 2.2 has a range of integration that is over all possible values of θ, and that it is a weighted average distribution, with the prior distribution $\pi_0(\theta)$ acting as the weighting function.

In cases where X is a discrete random variable (e.g., number of events in some period of time), $f(x)$ is the probability of seeing exactly x events, unconditional upon a value of θ. If X is a continuous outcome, such as time to suppress a fire, $f(x)$ is a density function, giving the unconditional probability of observing values of X in an infinitesimal interval about x. In a later context, associated with model validation, $f(x)$ will be referred to as the *predictive distribution* for X.

The likelihood function $f(x|\theta)$, or just likelihood, is also known by another name in PRA applications—it is the aleatory model describing an observed physical process. For example, a component failure to operate may be modeled inside a system fault tree by a Poisson process. Or, we may use an exponential distribution to represent fire suppression times. In these cases, there is an inherent modeling tie from the PRA to the data collection and evaluation process—specific aleatory models imply specific types of data. In traditional PRA applications, the aleatory model is most often binomial, Poisson, or exponential, giving rise to data in the form of failures over a specified number of demands, failures over a specified period of time, and failure times, respectively. Bayesian inference for each of these cases will be discussed in detail in Chap. 3.

The prior distribution, $\pi_0(\theta)$, represents what is known about a parameter θ independent of data generated by the aleatory model that will be collected and evaluated. Prior distributions can be classified broadly as either *informative* or *noninformative*. Informative priors, as the name suggests, contain substantive information about the possible values of θ. Noninformative priors, on the other hand, are intended to let the data dominate the posterior distribution; thus, they contain little substantive information about the parameter of interest.

2.3 A Simple Application of Bayes' Theorem

2.3.1 The Discrete Case

If we toss a coin, can we tell if it is an unfair coin? Specifically, what can the Bayesian approach to inference do to assist in answering this question? The issue that we are concerned with is the possibility of an unfair coin (e.g., a two-headed coin; for now, we will ignore the possibility of a two-tailed coin or a biased coin-tosser to simplify the presentation) being used. Let us jump directly into the Bayes analysis to see how straightforward this type of analysis can be in practice.

First, we note that Bayesian methods rely on three items:

- An aleatory model.
- A prior distribution for the parameter(s) of the aleatory model.
- Data associated with the aleatory model.

As discussed earlier, the prior distribution encodes the analyst's state of knowledge about a hypothesis. In this example, we have two hypotheses (**H**) we are going to consider:

H_1 = we have a fair coin
H_2 = we have an unfair coin

Recall in this example, that an unfair coin implies a two-headed coin. Thus, the probability of heads associated with H_2 would be 1.0 (since we cannot obtain a tail if in fact we have two heads). At this point, we are ready to specify the prior distribution.

Step 1: The Aleatory Model The likelihood function (or aleatory model) representing "random" outcomes (head/tail) for tossing a coin will be assumed to be given by a Bernoulli model:

$$P(D|H_i) = p_i$$

where p_i is the probability of obtaining a head on a single toss conditional upon the ith hypothesis.

Table 2.2 Bayesian inference of one toss of a coin in an experiment to test the hypothesis of a fair coin

Hypothesis	Prior probability	Likelihood	Prior × likelihood	Posterior probability
H_1: fair coin (i.e., the probability of a heads is 0.5)	0.75	0.5	0.375	0.60
H_2: two-headed coin (i.e., the probability of a heads is 1.0)	0.25	1.0	0.250	0.40
	Sum: 1.00		Sum: 0.625	Sum: 1.00

Step 2: The Prior Distribution Knowledge of the "experiment" might lead us to believe there is a significant chance that an unfair coin will be used in the toss. Thus, for the sake of example, let us assume that we assign the following prior probabilities to the two hypotheses:

$$P(H_1) = 0.75 \qquad P(H_2) = 0.25$$

This prior distribution implies that we think there is a 25% chance that an unfair coin will be used for the next toss. Expressed another way, this prior belief corresponds to odds of 3:1 that the coin is fair.

Step 3: The Data The coin is tossed once, and it comes up heads.

Step 4: Bayesian Calculation to Estimate Probability of a Fair Coin, $P(H_1)$ The normalization constant in Bayes' Theorem, $P(D)$, is found by summing the product of the prior distribution and the aleatory model over all possible hypotheses, which in this example gives

$$P(D) = P(H_1)p_1 + P(H_2)p_2 = (0.75)0.5 + (0.25)1.0 = 0.625$$

where for hypothesis $\mathbf{H_1}$, $p_1 = 0.5$ while for $\mathbf{H_2}$, $p_2 = 1.0$. At this point, we have the aleatory model (as a function of our one data point), the prior distribution, and the normalization constant in Bayes' Theorem. Thus, we can compute the posterior probabilities for our two hypotheses. When we do that calculation, we find:

$$P(\mathbf{H_1}| \text{ one toss, data are ``heads''}) = 0.6$$
$$P(\mathbf{H_2}| \text{ onetoss, data are ``heads''}) = 0.4$$

The results after one toss are presented in Table 2.2 and show that the posterior probability is the normalized product of the prior probability and the likelihood (e.g., H_1 posterior is $0.375 / 0.625 = 0.60$).

What has happened in this case is that the probability of the second hypothesis (two-headed coin) being true has increased by almost a factor of two simply by tossing the coin once and observing heads as the outcome.

As additional data are collected, we can evaluate the impact of the data on our state of knowledge by applying Bayes' Theorem sequentially as the data are

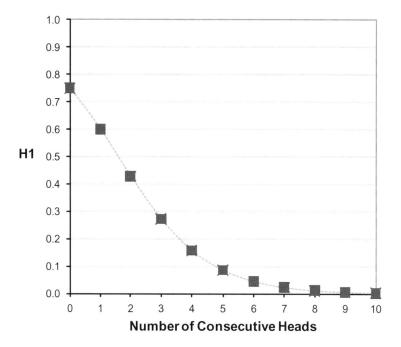

Fig. 2.1 Plot of the posterior probability of a fair coin as a function of the number of consecutive heads observed in independent tosses of the coin

collected. For example, let us assume that we toss the coin j times and want to make inference on the hypotheses (if a head comes up) each time. Thus, we toss $(x = 1, 2, \ldots, j)$ the coin again and again independently, and each time the estimate of the probability that the coin is fair changes. We see this probability plotted in Fig. 2.1, where initially (before any tosses) the prior probability of a fair coin (H_1) was 0.75. However, after five tosses where a head appears each time, the probability that we have a fair coin is small, less than ten percent.

2.3.2 The Continuous Case

Let us revisit the example described in Sect. 2.3.1 but employ a continuous prior distribution. The posterior distribution for this example will then be given by:

$$\pi_1(p \,|\, x, n) = \frac{f(x \,|\, p, n)\, \pi_0(\theta)}{\int_0^1 f(x \,|\, p, n)\, \pi_0(\theta)\, d\theta}$$

Since we used a Bernoulli aleatory model for the outcome of a coin toss, this leads to a binomial distribution as the aleatory model for the number of heads in n independent coin tosses in which the probability of heads on any toss is p:

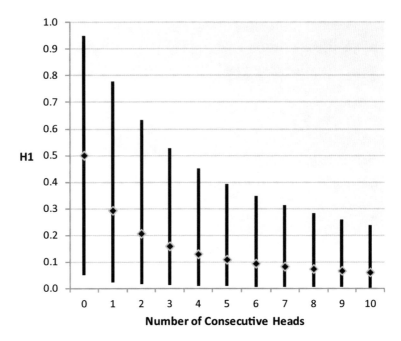

Number of Consecutive Heads

Fig. 2.2 Plot of the posterior probability of a fair coin as a function of the number of consecutive heads observed in independent tosses

$$f(x|p, n) = \binom{n}{x} p^x (1 - p)^{n-x}$$

To represent a prior state of ignorance, the prior could be specified as a uniform distribution between $p = 0$ and $p = 1$, or:

$$\pi_0(p) = 1 \qquad (0 \leq p \leq 1)$$

This is an example from the general category of noninformative priors mentioned earlier. We can now rewrite Bayes' Theorem for this example as:

$$\pi_1(p|x, n) = \frac{\theta^x (1 - \theta)^{n-x}}{\int_0^1 \theta^x (1 - \theta)^{n-x} d\theta} = \frac{\theta^x (1 - \theta)^{n-x}}{\frac{\Gamma(x+1)\Gamma(n-x+1)}{\Gamma(n+2)}}$$

It can be shown that the posterior distribution is a beta distribution with parameters:[1]

---

[1] The uniform distribution in this example is a particular instance of a conjugate prior, to be discussed in more detail in Chap. 3.

$$\alpha = x + 1$$
$$\beta = n - x + 1$$

Figure 2.2 shows the continuous case version of Fig. 2.1, where at each number of consecutive heads, the posterior beta distribution 5, 50 and 95th percentiles are displayed. For example, the prior density is initially uniform from 0 to 1, its 5th percentile is 0.05, its 50th is 0.5, and its 95th is 0.95. After three consecutive heads are observed, the posterior probability for the hypothesis H_1 (i.e., a fair coin) decreases: the 5th percentile of the posterior distribution for H_1 is 0.013, the 50th is 0.16, and the 95th is 0.53.

Chapter 3
Bayesian Inference for Common Aleatory Models

3.1 Introduction

We begin this chapter with the most commonly encountered modeling situations in risk assessment, which satisfy the following three conditions:

- The aleatory model contains a single unknown parameter,
- The prior information for this parameter is homogeneous and is known with certainty,
- The observed data are homogeneous and are known with certainty.

By homogeneous, we mean a set of information that is made up of similar constituents. A homogeneous population is one in which each item is of the same type. For example, a homogeneous population might be the four tires on an automobile (assuming proper tire rotation to equalize wear), or the tires in a fleet of automobiles of similar make, use, etc. Different types of tires, for example a tire from a small commuter automobile, may have different performance as compared to a tire from a large commercial truck. Inference for populations that are inhomogeneous is addressed in a later chapter. Inference for more complicated models, which contain more than a single unknown parameter, is described in later chapters.

3.2 The Binomial Distribution

The binomial distribution is often used as an aleatory model when a component must change state in response to a demand. For example, a relief valve may need to open to relieve pressure upon receipt of a signal from a controller that an over-pressure condition exists. The following assumptions underlie the binomial distribution:

D. Kelly and C. Smith, *Bayesian Inference for Probabilistic Risk Assessment*,
Springer Series in Reliability Engineering, DOI: 10.1007/978-1-84996-187-5_3,
© Springer-Verlag London Limited 2011

1. There are two possible outcomes of each demand, typically denoted by success and failure.
2. There is a constant probability (p) of failure (or success) on each demand.
3. The outcomes of each demand are independent, that is, earlier demands do not influence the outcomes of later demands (i.e., the order of failures/successes is irrelevant).

The unknown parameter in this model is p, and the observed data are the number of failures, denoted by x, in a specified (i.e., known) number of demands, denoted by n. Both x and n are assumed to be known with certainty in this chapter. Cases in which x and n are uncertain are treated in a later chapter.

The probability for obtaining x failures in n demands is given by the binomial (p, n) distribution as

$$Pr(X = x) = f(x|p) = \binom{n}{x} p^x (1 - p)^{n-x}$$

where $0 \leq x \leq n$ and $\binom{n}{x}$ is the binomial coefficient. The binomial coefficient gives the number of ways that x failures can occur in n demands (i.e., the number of combinations of n demands selected x at a time).

Note that the binomial distribution describes the aleatory uncertainty in the observed number of failures, x. Bayesian inference describes how the epistemic uncertainty in p changes from that encoded in the prior distribution, describing the analyst's state of knowledge about possible values of p before x is observed, to the posterior distribution, which reflects how the data have altered the analyst's state of knowledge about p.

3.2.1 Binomial Inference with Conjugate Prior

The simplest type of prior distribution from the standpoint of the mathematics of Bayesian inference is a so-called conjugate prior, in which the prior and posterior distribution are of the same functional type (e.g., beta, gamma), and the integration needed to obtain the normalizing constant in Bayes' Theorem is effectively circumvented. Not every aleatory model will have an associated conjugate prior, and we may sometimes choose to use a nonconjugate prior even when a conjugate prior is available, but the three most commonly used aleatory models in PRA have associated conjugate priors. For the binomial distribution, the conjugate prior is a beta distribution.

Two parameters are needed to specify the beta prior distribution, and these will be denoted α_{prior} and β_{prior}. Conceptually, α_{prior} can be thought of as the number of failures contained in the prior distribution, and the sum of α_{prior} and β_{prior} is like the number of demands over which these failures occurred. Thus, small values of α_{prior} and β_{prior} correspond to less information, and this translates into a broader, more diffuse prior distribution.

With the data consisting of x failures in n demands, the conjugate nature of the prior distribution and likelihood function allows the posterior distribution to be determined using arithmetic. The posterior distribution is also a beta distribution, with adjusted (labeled "post") parameters given by:

$\alpha_{post} = \alpha_{prior} + x$
$\beta_{post} = \beta_{prior} + n - x$.

From the properties of the beta distribution, the prior and posterior mean of p are given by:

Prior mean $= \alpha_{prior}/(\alpha_{prior} + \beta_{prior})$
Posterior mean $= \alpha_{post}/(\alpha_{post} + \beta_{post}) = (\alpha_{prior} + x)/(\alpha_{prior} + \beta_{prior} + n)$

Credible intervals for either the prior or the posterior must be found numerically, and this can be done using the BETAINV() function built into a spreadsheet program.

Numerically simulating Bayes Theorem is possible using MCMC (via tools like OpenBUGS, to be discussed later) with the applicable prior distribution and aleatory model specified. Note that when using a conjugate prior, numerical simulation is not needed since the posterior can be found directly, but numerical simulation is a general method for Bayesian inference.

As an example, assume that we are going to model a power supply failing to energize. We will use a binomial aleatory model and assume a conjugate prior distribution for p. The prior distribution for failure of the power supply on demand is given (from an industry database) as a beta distribution with parameters

$\alpha_{prior} = 1.24$
$\beta_{prior} = 189,075$.

Further, assume 2 failures to function have been seen in 285 demands on the power supply. Now that we have an aleatory model, a prior distribution on the parameter of the model, and associated data, we can perform Bayesian inference. Specifically, we can find the posterior mean of p, the probability that the power supply fails to function on demand, and a 90% credible interval for p, which summarizes the epistemic uncertainty encoded in the posterior beta distribution for p.

We begin by noting that the mean of the beta prior distribution is $1.24/(1.24 + 189,075) = 6.56 \times 10^{-6}$. The prior distribution expresses significant epistemic uncertainty about the value of p. This can be seen by calculating a 90% prior credible interval for p. We can use the BETAINV() function in a spreadsheet to do this, as mentioned above. The 5th percentile of the prior distribution is given by BETAINV(0.05, 1.24, 189075) $= 5.4 \times 10^{-7}$ and the 95th percentile is given by BETAINV(0.95, 1.24, 189075) $= 1.8 \times 10^{-5}$, an uncertainty range of almost two orders of magnitude.

With 2 failures of the power supply in 285 demands, along with the assumption that these failures are described adequately by a binomial distribution, the posterior distribution is also a beta distribution, with parameters $\alpha_{post} = 1.24 + 2 = 3.24$ and $\beta_{post} = 189,075 + 285 - 2 = 189,358$. The posterior mean of p is given by $3.24/(3.24 + 189,358) = 1.7 \times 10^{-5}$. The 90% posterior credible

Fig. 3.1 Comparison
of prior and posterior
distributions for the
evaluation of the power
supply

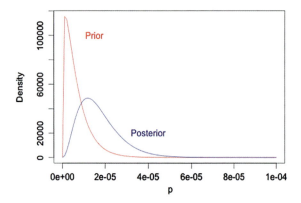

Table 3.1 OpenBUGS script implementing a conjugate beta prior distribution for a binomial aleatory model

model {	# A Model is defined between { } symbols
x ~ dbin(p, n)	# Binomial dist. for number of failures in n demands
p ~ dbeta(alpha.prior, beta.prior)	# Conjugate beta prior distribution for p
}	
data	
list(x = 2, n = 285)	# Power supply failure data
list(alpha.prior = 1.24,	# Prior parameters
beta.prior = 189075)	

interval is found using the BETAINV() function, just as was done for the prior interval above. The posterior 5th percentile is given by BETAINV(0.05, 3.24, 189358) = 5.0×10^{-6} and the 95th percentile is given by BETAINV(0.95, 3.24, 189358) = 3.5×10^{-5}. Note how the epistemic uncertainty in the prior distribution has been reduced by the observed data. This is shown graphically in Fig. 3.1, which overlays the prior and posterior distribution for this example.

Inference for conjugate cases like the power supply example can also be carried out using MCMC approaches (such as with OpenBUGS). Table 3.1 shows the implementation of the example for the binomial/beta conjugate example just covered. For problems such as this, where there is only one unknown parameter to be estimated, so the sampling is very fast, we advise the analyst to use at least 100,000 iterations, discarding the first 1,000 to allow for convergence to the posterior distribution (convergence checks are discussed in a later chapter). Monitoring node p will display the desired posterior results.

Within the table below (and the remaining OpenBUGS examples in this and later chapters), the following notation is used:

" ~ " indicates that the variable to the left of " ~ " is described by the distribution on the right of " ~ ." Examples of distributions we will use commonly include binomial (dbin), beta (dbeta), gamma (dgamma), Poisson (dpois), normal (dnorm), and lognormal (dlnorm). Other distributions will be encountered less

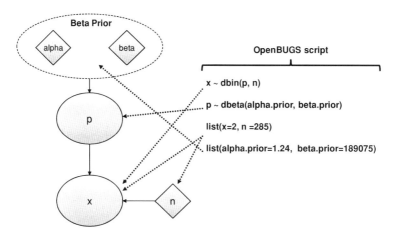

Fig. 3.2 DAG representing the example in Table 3.1

frequently, and will be introduced and explained as needed. The built-in Help feature for OpenBUGS gives details on the distributions that can be used. A summary of distribution properties can be found in Appendix A.

"#" indicates that the text to the right of "#" is a comment.

"<-" indicates that the variable to the left of "<-" is equivalent to the expression on the right of "<-."

A Bayesian network or directed acyclic graph (DAG) is a common way of displaying a Bayesian inference problem and is the underlying model used by OpenBUGS (in script form). In a DAG, observed aleatory variables (also referred to as *nodes*),(such as *x* in the binomial inference problem above, are displayed as ovals that contain no child nodes (i.e., they occur at the "lowest" level in the diagram). Uncertain variables that influence *x* are shown as ovals at a higher level in the DAG, and are connected by arrows to the variables they influence (i.e., they are parents of the nodes they influence). Constant parameters (such as *n* in the table above) are sometimes shown in the DAG as rectangles or diamonds.

For relatively simple problems, a DAG can be an aid in understanding the problem, particularly for an analyst who is new to OpenBUGS. However, as the complexity of the problem increases, most experienced analysts will find that the script representation of the problem is often clearer.

Figure 3.2 shows the DAG corresponding to the OpenBUGS script found in Table 3.1 (where specific lines of script are noted). This DAG illustrates that *x* is the observed variable, because it is a node with no children. This node (*x*) is an uncertain variable, indicated by its oval shape. Its value is influenced by *p* (*p* is a parent node to *x*), which is the parameter of interest in this problem; we observe *x* (with *n* specified), and use this information to infer possible values for *p*. The dashed region at the top of the DAG, labeled "Beta Prior," clarifies the type of prior distribution used for *p*, and indicates that the parameters of this distribution (*alpha* and *beta*) are specified by the analyst.

3.2.2 Binomial Inference with Noninformative Prior

As the name suggests, a noninformative prior distribution contains little information about the parameter of interest, which in this case is p. Such priors originated in a (continuing) quest to find a mathematical representation of complete uncertainty. This has led some to conclude that noninformative priors should be used when one knows nothing about the parameter being estimated. As discussed in earlier sections, this is almost never the case in practice, and use of a noninformative prior in such a case can lead to excessively conservative results. There are two situations in which a noninformative prior may be useful:

1. The first is where the observed data are abundant enough to dominate the information contained in any reasonable prior, so it is not worthwhile to expend resources developing an informative prior distribution.
2. The second is where the analyst wishes to use a prior that has little influence on the posterior, perhaps as a point of reference, or to illustrate the impact of an alternative informative prior.

With an abundance of observed data, the prior distribution will have little influence on the posterior. So why not just use the data alone? Remember that Bayesian inference uses probability distributions to represent uncertainty, and more pragmatically, a probability distribution is needed to propagate uncertainty through the PRA models, so Bayes' Theorem is still used to obtain a posterior distribution.

Noninformative priors are also known as formal priors, reference priors, diffuse priors, objective priors, and vague priors. For clarity, we will refer to them as noninformative priors in this text, as that is the name most commonly employed in PRA. Unfortunately, because of the difficulty in representing a state of ignorance with a probability distribution, there are many routes that lead to slightly different noninformative priors, and the intuitive choice of a uniform prior, which gives equal weight to all possible subintervals of equal length in the parameter range, is not what is usually used. The most common noninformative prior for single-parameter inference in PRA is the "Jeffreys prior."

The Jeffreys functional form is dependent upon the likelihood function, so there is not a single "Jeffreys prior" for all cases. Instead, there is a different Jeffreys prior for each likelihood function (aleatory model). For the case here, where the likelihood function is the binomial distribution, the Jeffreys prior is a beta distribution with both parameters equal to 0.5. Thus, from the perspective of the mathematics of obtaining the posterior distribution, inference with the Jeffreys prior is a special case of inference with a beta conjugate prior. Using the Jeffreys prior with the binomial aleatory model leads to a beta posterior distribution with parameters $x + 0.5$ and $n - x + 0.5$, and a posterior mean of $(x + 0.5)/(n + 1)$.

Note that if x and n are small (i.e., sparse data), then adding "half a failure" to x may give a result that is felt to be too conservative. In such cases, a possible alternative noninformative prior to the Jeffreys prior is like a beta distribution with both parameters equal to zero (the "zero–zero" beta distribution). This is not a proper probability distribution, but as long as x and n are both greater than zero, the posterior distribution will be proper and the posterior mean will be x/n.[1]

Conceptually, adjusting the beta prior so that α_{prior} and α_{prior} both have small values (in the limit, zero) tends to reduce the impact of the prior on the posterior mean and allows the data to dominate the results. Note, though, that when α_{prior} and β_{prior} are equal and not zero, the mean of this beta prior is 0.5. The prior should reflect what information, if any, is known independent of the data.

3.2.3 Binomial Inference with Nonconjugate Prior

A nonconjugate prior is one in which the prior and posterior distribution are not of the same functional form. In such cases, numerical integration is required to find the normalizing constant in the denominator of Bayes' Theorem. In the past, this requirement of integration has been a computational limitation of Bayesian inference, and is one reason for the popularity of conjugate priors. However, cases often arise in which a nonconjugate prior is desirable, despite the increased mathematical difficulty. As an example, generic databases often express epistemic uncertainty in terms of a lognormal distribution, which is not conjugate with the binomial likelihood function. Additionally, conjugate priors have relatively light tails and can overly influence the results in cases where there is sparse data that is in conflict with the prior. The estimate provided by the data will typically lie in the tail of the prior in such cases, where the prior probability is very small. In this section, we describe how to carry out inference with a lognormal prior, which is a commonly-encountered nonconjugate prior, and with a logistic-normal prior, which is similar to a lognormal prior but is more appropriate when the values of p are expected to be nearer one.[2]

Although spreadsheets can be used to carry out the required numerical integration for the case of a single unknown parameter, another way to implement nonconjugate priors is with OpenBUGS. In this example, we will assume that we are using a binomial aleatory model to model failure to open of a valve in a pipe. Further, assume that instead of a conjugate prior (as in the previous example), we are using a generic database that provides a lognormal prior for p, where the database lists the mean failure probability as 10^{-6} with a lognormal "error factor" of 10 (note that the error factor is defined as either of the

---

[1] Because the posterior credible intervals obtained from updating the "zero–zero" prior do not provide adequate coverage, we do not advocate the routine use of this prior as a replacement for the Jeffreys prior. The zero–zero prior is systematically biased low, in the anti-conservative direction. A Bayesian making supposedly fair bets would not come out even in the long run. Also, choosing a prior *after* examining the data violates the spirit of Bayesian inference.

[2] Note that in a *reliability* analysis, as opposed to a PRA, the unknown parameter of interest might be $q = 1-p$, in which case q would be close to one for highly reliable components.

Table 3.2 OpenBUGS script for Bayesian inference with binomial likelihood function and a lognormal prior

```
model {
x ~ dbin(p, n)                              # Binomial model for number of failures
p ~ dlnorm(mu, tau)                         # Lognormal prior distribution for p
tau <- 1/pow(log(prior.EF)/1.645, 2)        # Calculate tau from lognormal error
                                              factor

# Calculate mu from lognormal prior mean and error
    factor
mu <- log(prior.mean) - pow(log(prior.EF)/1.645, 2)/
    2
}
data
list(x = 2, n = 285, prior.mean = 1.E-6,
    prior.EF = 10)
```

Fig. 3.3 Mean and 90% credible interval for the script shown in Table 3.2

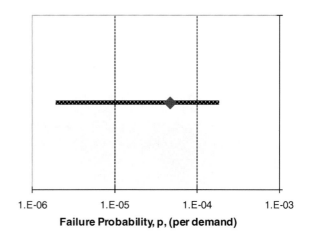

Failure Probability, p, (per demand)

1.E-06   1.E-05   1.E-04   1.E-03

following for the lognormal distribution: the ratio of the 95th percentile to the 50th, the ratio of the 50th percentile to the 5th, or the square root of the ratio of the 95th percentile to the 5th). Assume that our observed data are 2 failures in 285 demands. The OpenBUGS script shown in Table 3.2 is used to analyze this example.

Running he script in Table 3.2 for 100,000 iterations, after discarding the first 1,000 iterations to allow for convergence to the posterior distribution, gives a posterior mean for p of 4.7×10^{-5} and a 90% credible interval of (1.9×10^{-6}, 1.8×10^{-4}) (see Fig. 3.3). Note that when the prior distribution is not conjugate, the posterior distribution cannot be written down in closed form. In such cases, an analyst may replace the numerically defined posterior with a distribution of a particular functional form (e.g., lognormal), or may use the empirical results of the OpenBUGS analysis to construct a histogram.

Table 3.3 OpenBUGS script for Bayesian inference with binomial likelihood function and logistic-normal prior

model {	
x ~ dbin(p, n)	# Binomial distribution for number of failures
p <- exp(p.constr)/(1 + exp(p.constr))	# Logistic-normal prior distribution for p
p.constr ~ dnorm(mu, tau)	
tau <- pow(sigma, -2)	
sigma <-(log(p.95/(1− p.95)) − mu)/1.645	
# Calculate mu from lognormal median	
mu <- log(prior.median/(1− prior.median))	
p.95 <- prior.median*prior.EF	
}	
data	
list(x = 2,n = 256, prior.median = 1.E-6,	
prior.EF = 10)	

Generic databases may not always describe the lognormal distribution in terms of a mean value and an error factor; quite often the median (50th percentile) is specified rather than the mean value. This may also be the case when eliciting information from experts as an expert may be more comfortable providing a median value. In this case, the analysis changes only slightly. For example, in the script provided in Table 3.2, the line that calculates mu from the lognormal prior mean and error factor is replaced by the following line:

mu <-log(prior.median)

and prior.median is loaded in the data statement instead of prior.mean.

Note that in the previous example we were using a lognormal distribution to represent uncertainty in a probability. A potential problem with this representation is that the lognormal distribution can have values greater than one, and as such may not faithfully represent uncertainty in a parameter that should be constrained to be less than one. Cases may arise where the value of p could be approaching unity. In such cases, a logistic-normal prior is a "lognormal-like" distribution, but one that constrains the values of p to lie between zero and one. The OpenBUGS script shown in Table 3.3 uses the lognormal median and error factor (e.g., from a generic database), but "constrains" the distribution to lie between zero and one by replacing the lognormal distribution with a logistic-normal distribution.[3]

---

[3] Note that the parameters of the logistic-normal distribution are not related to the mean and error factor in as simple a fashion as was the case for the lognormal distribution; the mean and higher moments of the logistic-normal distribution must be calculated numerically. Thus, for simplicity, we use the median and error factor, for which the relations are algebraic, as shown in the script.

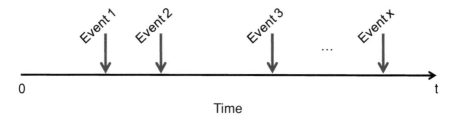

Fig. 3.4 Representation of Poisson-distributed events in time

3.3 The Poisson Model

The Poisson model is often used in PRA for failures of normally operating components, failures of standby components that occur at some point in time prior to a demand for the component to change state, and for initiating events. The following assumptions underlie the Poisson distribution:

- The probability of an event (e.g., a failure) in a small time interval is approximately proportional to the length of the interval. The constant of proportionality is denoted by *lambda* (λ).
- The probability of simultaneous events in a short interval of time is approximately zero.
- The occurrence of an event in one time interval does not affect the probability of occurrence in another, non-overlapping time interval.

Note that λ represents a rate and has units of inverse time. Also note that λ is not a function of time, so the simple Poisson distribution cannot be used for reliability growth or aging (see Chaps. 8 and 9 for these topics).

The unknown parameter in this model is λ, and the observed data are the number of events, denoted by x, in a specified time period, denoted by t. The form of the data for the Poisson model is shown graphically in Fig. 3.4. Both x and t are assumed to be known with certainty in this chapter. Cases in which x and t may also have epistemic uncertainty are treated in Chap. 10.

The probability for observing x events in time t is given by the Poisson(λt) distribution as

$$Pr(X = x) = f(x|\lambda) = \frac{(\lambda t)^x e^{-\lambda t}}{x!}$$

where $x = 0, 1, 2,\ldots$ and λ is the rate of the Poisson process.

The Poisson model is used to describe the aleatory uncertainty in the number of events, x. In other words, we cannot predict exactly how many Poisson events will be seen over some period of time. Bayesian inference is then used to describe how the epistemic uncertainty in λ changes from the prior distribution, which describes the analyst's state of knowledge about possible values of λ before empirical data are collected, to the posterior distribution, which reflects how the observed data have altered the analyst's prior state of knowledge about possible values of λ.

3.3.1 Poisson Inference with Conjugate Prior

As was the case with the binomial distribution, a conjugate prior is sometimes chosen for purposes of mathematical convenience. For the Poisson distribution, the conjugate prior is a gamma distribution.

As was the case for the beta distribution, two parameters are needed to specify a gamma distribution, and these are denoted α_{prior} and β_{prior}. Do not confuse α_{prior} and β_{prior} here with the parameters of the beta distribution in the previous section; here they represent the parameters of a gamma distribution.

Conceptually, α_{prior} can be thought of as the number of events contained in the prior distribution, and β_{prior} is like the period of time over which these events occurred. Thus, small values of α_{prior} and β_{prior} correspond to little information, and this translates into a broader, more diffuse prior distribution for λ. Also note that β_{prior} has units of time, and the units have to be the same as the units of t.

With the observed data consisting of x failures in time t, the conjugate nature of the prior distribution and likelihood function allows the posterior distribution to be written down immediately using simple arithmetic. The posterior distribution is also a gamma distribution, with new (adjusted) parameters given by:

$$\alpha_{post} = \alpha_{prior} + x$$
$$\beta_{post} = \beta_{prior} + t.$$

From the properties of the gamma distribution (see Appendix B), the prior and posterior mean of λ are given by $\alpha_{prior}/\beta_{prior}$ and $\alpha_{post}/\beta_{post}$, respectively. Credible intervals for either distribution can be found using the GAMMAINV() function built into modern spreadsheets. Caution: Be sure to know how your spreadsheet software parameterizes the gamma distribution. Many spreadsheets use the reciprocal of β as the scale parameter.

As an example of the conjugate calculation for the Poisson aleatory model, assume we are representing an air fan failing to operate (Poisson model) with a gamma prior with parameters $\alpha_{prior} = 1.6$ and $\beta_{prior} = 365,000$ h used to represent uncertainty in the fan failure rate, λ. No failures are observed in 200 days of operation of the fan. We can now find the posterior mean and 90% credible interval for the air fan failure rate.

Because the gamma prior distribution is conjugate to the Poisson likelihood function, the posterior distribution will also be a gamma distribution, with parameters $\alpha_{post} = 1.6 + 0 = 1.6$ and $\beta_{post} = 365,000$ h + (200 days) (24 h/day) = 369,800 h. The posterior mean is the ratio of α_{post} to β_{post}, which is 4.3×10^{-6}/h. The 90% credible interval is found using the GAMMAINV() function in a spreadsheet. As noted above, most spreadsheet software uses the reciprocal of β as the second parameter. This can be dealt with either by entering $1/\beta$ as the second argument, or by entering 1 as the second parameter and dividing the result by β. Thus, the 5th percentile is given by = GAMMAINV(0.05, 1.6, 1/369800), which produces the result 5.6×10^{-7}/h. Similarly, the 95th percentile is given by = GAMMAINV(0.95, 1.6, 1/369800) which gives 1.1×10^{-5}/h.

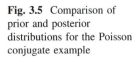

Fig. 3.5 Comparison of prior and posterior distributions for the Poisson conjugate example

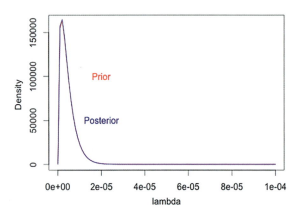

Table 3.4 OpenBUGS script for Bayesian inference with Poisson likelihood and gamma conjugate prior

```
model {
x ~ dpois(mean.poisson)              # Poisson likelihood function
mean.poisson <-lambda*time.hr        # Parameterize in terms of failure rate, lambda
time.hr <-time*24                    # Convert days to hours
lambda ~ dgamma(1.6, 365000)         # Gamma prior for lambda
}
data
list(x = 0, time = 200)
```

We can now use the posterior mean to find the probability that the air fan will operate successfully for a mission time of 1,000 h (a reliability question). Using the posterior mean failure rate of 4.3×10^{-6}/h, the probability that the air fan operates successfully for 1,000 h is just $\exp[-(4.33 \times 10^{-6}/\text{h})(1{,}000 \text{ h})] = 0.996$.

The plot in Fig. 3.5 shows how little the prior distribution has been affected by the relatively sparse data in this example (the posterior distributions is mostly on top of the prior distribution).

Bayesian inference for this type of problem can also be carried out using OpenBUGS. The script shown in Table 3.4 implements the analysis. OpenBUGS, unlike most spreadsheets, uses β rather than $1/\beta$ to parameterize the gamma distribution. Monitoring node lambda will display the desired posterior results (Fig. 3.6).

3.3.2 Poisson Inference with Noninformative Prior

As was the case for the binomial distribution, there are many routes to a noninformative prior for λ in the Poisson model. As mentioned before, the most commonly used noninformative prior in PRA is the Jeffreys prior. In the case of the

Fig. 3.6 DAG for the
Poisson conjugate calculation
shown in Table 3.4

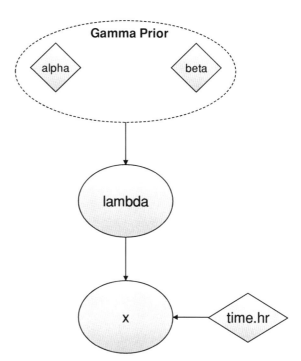

Poisson likelihood the Jeffreys noninformative prior is like a gamma distribution
with $\alpha_{\text{prior}} = 0.5$ and $\beta_{\text{prior}} = 0$. This is not a proper distribution, as the integral
over all possible values of λ is not finite. However, it always yields a proper
posterior distribution, with parameters $\alpha_{\text{post}} = x + 0.5$ and $\beta_{\text{post}} = t$. Thus, the
posterior mean of lambda is given by $(x + 0.5)/t$. Inference with the Jeffreys prior
can be thought of as a special case of inference with a gamma conjugate prior.

Note that if x and t are small (sparse data), then adding "half an event" to x may give
a result that is felt to be too conservative. In such cases, a possible alternative to the
Jeffreys prior is like a gamma distribution with both parameters equal to zero. This is
not a proper probability distribution, but as long as x and t are greater than zero, the
posterior distribution will be proper and the posterior mean will take on the value (x/t).[4]

3.3.3 Poisson Inference with Nonconjugate Prior

As was the case for the parameter p in the binomial distribution, a lognormal
distribution is a commonly encountered nonconjugate prior for λ in the Poisson
distribution. The analysis can be carried out with OpenBUGS, exactly as was done

---

[4] As discussed for the binomial distribution earlier, we do not advocate the routine use of the
"zero–zero" gamma prior as a replacement for the Jeffreys prior.

Table 3.5 OpenBUGS script for Bayesian inference with Poisson likelihood function and lognormal prior

model {	
x ∼ dpois(mean.poisson)	# Poisson distribution for number of events
mean.poisson <-lambda*time.hr	# Poisson parameter
time.hr <-time*24	# Convert days to hours
lambda ∼ dlnorm(mu, tau)	# Lognormal prior distribution for lambda
tau <-1/pow(log(prior.EF)/1.645, 2)	# Calculate tau from lognormal error factor
mu <-log(prior.median)	# Calculate mu from lognormal median
}	
data	
list(x = 0, time = 200, prior.median = 5.E-7, prior.EF = 14)	

for p in the binomial distribution. Here, however, there is no concern about values of λ greater than one, because λ is a rate instead of a probability, and can take on any positive value, in principle.

As an example of the Poisson nonconjugate case, assume we are modeling a pump failing to operate (Poisson aleatory model) and assume that the prior distribution describing the uncertainty in the failure rate of the circulating pump is lognormal with a median of 5×10^{-7}/h and an error factor of 14. Assume the observed data are no failures in 200 days. The OpenBUGS script in Table 3.5 can be used to find the posterior mean and 90% credible interval for λ.

Running this script for 100,000 iterations, after discarding the first 1,000 iterations to allow for convergence, gives a posterior mean for λ of 1.6×10^{-6}/h and a 90% credible interval of (3.5×10^{-8}/h, 6.5×10^{-6}/h).

3.4 The Exponential Model

There are cases where we observe the times at which random events occur instead of the number of such events in a specified period of time. Examples are times to failures of components, times to suppress a fire, etc. If the assumptions for the Poisson distribution listed earlier in this chapter (restated below) are met, then the times between events will be exponentially distributed with unknown parameter λ; this is the same λ that appears as the unknown parameter in the Poisson distribution. Thus, if the *times* at which Poisson-distributed events occur are observed, then the likelihood function is now based on the exponential distribution. The following assumptions underlie the exponential distribution:

• The probability of an event (e.g., a failure) in a small time interval is approximately proportional to the length of the interval. The constant of proportionality is denoted by λ.

Fig. 3.7 Representation of exponentially distributed events in time

- The probability of simultaneous events in a short interval of time is approximately zero.
- The occurrence of an event in one time interval does not affect the probability of occurrence in another, non-overlapping time interval.
- The random event that is observed is the time to an event.

Because the observed data consist of n failure times (with n now specified rather than random), the form of the likelihood function changes from a Poisson distribution to a product of n exponential density functions. The concept of exponential data is illustrated in Fig. 3.7.

In this chapter, we treat only the case in which all failure times are observed and known with certainty. In later chapters, we will cover cases in which not all components fail (i.e., censored data), and in which observed failure times have epistemic uncertainty (i.e., the times are uncertain).

3.4.1 Exponential Inference with Conjugate Prior

As was the case for the Poisson distribution, the conjugate prior for the exponential likelihood is again a gamma distribution, with parameters denoted α_{prior} and β_{prior}. Once again, β_{prior} has units of time, and these units must match the units of the observed times that constitute the data. The posterior distribution will be a gamma distribution with parameters $\alpha_{\text{post}} = \alpha_{\text{prior}} + n$ (the number of observed times), and $\beta_{\text{post}} = \beta_{\text{prior}} + t_{\text{total}}$, where t_{total} is the sum of the observed times.

From the properties of the gamma distribution the prior and posterior mean of λ are given by $\alpha_{\text{prior}}/\beta_{\text{prior}}$ and $\alpha_{\text{post}}/\beta_{\text{post}}$, respectively. Credible intervals for either distribution can be found using the GAMMAINV() function built into modern spreadsheets, observing the earlier caution about parameterization of the gamma distribution.

As an example of the exponential conjugate calculation, we will model a pump failing to operate (exponential model) and use a gamma prior to describe uncertainty in the pump failure rate. Assume we have collected 7 times to failure (in hours) for pumps: 55707, 255092, 56776, 111646, 11358772, 875209 and 68978. If the prior distribution is gamma with $\alpha_{\text{prior}} = 1.6$ and $\beta_{\text{prior}} = 365,000$ h, we can find the posterior mean and 90% credible interval for the pump failure rate λ.

The posterior distribution is gamma with parameters $\alpha_{\text{post}} = 1.6 + 7 = 8.6$ and $\beta_{\text{post}} = 365,000 + 12,782,181 = 13,147,181$ h. The posterior mean is given by

Table 3.6 OpenBUGS script for Bayesian inference with an exponential likelihood and gamma conjugate prior

```
model        {
for(i in 1:n)               {
             time[i] ~ dexp(lambda)      # Exponential likelihood function for n failure times
             }
lambda ~ dgamma(alpha, beta)            # Gamma prior for lambda
}
data                                    # Note the nested () for the time array
list(time = c(55707, 255092, 56776, 111646, 11358772, 875209, 68978), n = 7, alpha = 1.6,
   beta = 365,000)
```

$\alpha_{post}/\beta_{post} = 6.5 \times 10^{-7}$/h. The 5th percentile is given by $= GAMMAINV(0.05, 8.6, 1/13147181)$ which is 3.4×10^{-7}/h. The 95th percentile is given by $= GAMMAINV(0.95, 8.6, 1/13147181)$ which is 1.1×10^{-6}/h.

OpenBUGS can also be used for this example, as shown in Table 3.6.

3.4.2 Exponential Inference with Noninformative Prior

The Jeffreys noninformative prior for the exponential likelihood is like a gamma distribution with both parameters equal to zero. This might seem odd, given the relationship between the exponential and Poisson distributions mentioned above. In fact, it is odd that the Jeffreys prior changes, depending on whether one counts failures or observes actual failure times. The Jeffreys prior in this case is an improper distribution, but it always results in a proper posterior distribution. The parameters of the posterior distribution will be n and t_{total}, resulting in a posterior mean of n/t_{total}. This mean is numerically equal to the frequentist maximum likelihood estimate (MLE), and credible intervals will be numerically equal to confidence intervals from a frequentist analysis of the data.

Note that OpenBUGS can only accept proper distributions, so the Jeffreys prior for exponential data is specified as dgamma(0.0001, 0.0001). An initial value for λ will have to be provided, as OpenBUGS cannot generate an initial value from this distribution.

3.4.3 Exponential Inference with Nonconjugate Prior

Again, the lognormal distribution is a commonly encountered nonconjugate prior for a failure rate. To perform this analysis, we will use OpenBUGS.

For the example, we will assume we are modeling a pump failing to operate (exponential model) with a lognormal prior describing the uncertainty in the pump failure rate. Assume the prior median is 5×10^{-7}/h with an error factor of 14. Also, we will use the failure times from the previous example as our observed

Table 3.7 OpenBUGS script for Bayesian inference with exponential likelihood function and lognormal prior

model {	
for(i in 1:n) {	
time[i] ~ dexp(lambda)	# Exponential likelihood function for n failure times
}	
lambda ~ dlnorm(mu, tau)	# Lognormal prior for lambda
tau <-1/pow(log(prior.EF)/1.645, 2)	# Calculate tau from lognormal error factor
mu <-log(prior.median)	# Calculate mu from lognormal mean
}	
data	# Note the nested () for the time array
list(time = c(55707, 255092, 56776, 111646, 11358772, 875209, 68978), n = 7,	
prior.median = 5.E-7, prior.EF = 14)	

data. We can now find the posterior mean and 90% interval for the failure rate using the script shown in Table 3.7.

Using 100,000 iterations, with 1,000 burn-in iterations discarded to allow for convergence to the posterior distribution, the posterior mean of λ is 5.5×10^{-7}/h, with a 90% credible interval of $(2.6 \times 10^{-7}$/h, 9.2×10^{-7}/h).

3.5 Developing Prior Distributions

In practice, the analyst must develop a prior distribution from available engineering and scientific information, where the prior should:

- Reflect what information is known about the inference problem at hand
- Be independent of the data that is collected.

This section provides some high-level guidance and examples for developing priors for single-parameter problems. In this section we consider only priors developed from a single source of information; later chapters will deal with developing priors from multiple sources of information.

3.5.1 Developing a Conjugate Prior

The beta and gamma distributions used as conjugate priors in the earlier sections each have two parameters that specify the distribution. Therefore, two independent pieces of information are generally needed to select such a conjugate prior. Common information from which the analyst can develop a prior includes:

- A measure of central tendency (e.g., median or mean) and an upper bound (e.g., 95th percentile)

- Upper and lower bound (e.g., 95th and 5th percentile)
- A mean and variance (or standard deviation).

We discuss each of these three cases below.

Using Mean or Median and Upper Bound—When the information provided[5] takes the form of a mean or median value and an upper bound, numerical analysis is required in order to find a gamma or beta distribution satisfying this information. Fortunately, modern spreadsheet tools make such analysis feasible. Note that "bound" is not usually interpreted in an absolute sense as a value that cannot be exceeded. Instead, it is interpreted as an upper percentile of the distribution. The 95th percentile is the most common choice in PRA, but other percentiles, such as the 85th, can be chosen.

As an example, assume we were provided information on a part failure rate such as:

- Median of 2×10^{-10}/h
- A 95th percentile that is ten times the point estimate (i.e., 2×10^{-9}/h).

To use this information to develop a gamma prior distribution for the failure rate, there are two equations (one for the median, one for the 95th percentile) that must be solved numerically to find α and β (the parameters of the gamma prior distribution).[6]

First, using the median value, we can note that:

=GAMMAINV(0.5, α, $1/\beta$) = 2×10^{-10}/h

Second, using the 95th percentile, we can note that:

=GAMMAINV(0.95, α, $1/\beta$) = 2×10^{-9}/h

To solve these equations, we could iterate on values of α and β until the equalities were satisfied to some numerical precision. Alternatively, one can set up a spreadsheet to allow use of the SOLVER function.

Using the SOLVER function (set the initial α and β values to 0.1 and 0.01 respectively and set the precision to 1E-10) yields a gamma distribution with α equal to about 0.6 and β equal to about 1.2×10^9 h.

Using Upper and Lower Bound—Sometimes information will be provided in the form of a range, from a lower bound to an upper bound. As above, the bounds are not typically absolute bounds on the parameter value, but are interpreted as lower and upper percentiles (e.g., 5th and 95th) of the prior distribution. Again, we will have to resort to numerical methods (e.g., the SOLVER function) to find the parameters of a conjugate prior distribution.

---

[5] The "information provided" represents the analyst's state of knowledge for the system or component being evaluated and must be independent from any data to be used in updating the prior distribution.

[6] We use a spreadsheet tool in what follows, but a more accurate alternative is the Parameter Solver software, developed by the M. D. Anderson Cancer Center. It can be downloaded free of charge from https://biostatistics.mdanderson.org/SoftwareDownload/ProductDownloadFiles/ParameterSolver_V2.3_WithFX1.1.exe

Using Mean and Variance or Standard Deviation—This is the easiest situation to deal with, but perhaps the least frequently encountered in practice. It is relatively easy because the equations are algebraic and do not require numerical solution. Unfortunately, our information sources are not often encoded in terms of the mean and variance or standard deviation (the square root of the variance).

In the past, some analysts, in an attempt to avoid working with a nonconjugate prior, have converted the encoded information into a mean and variance in order to replace the nonconjugate prior with a conjugate prior using simple algebraic calculations. This practice can lead to nonconservatively low estimates in some cases, however, and is not recommended in today's environment where tools such as OpenBUGS make Bayesian inference with a nonconjugate prior straightforward.

For the gamma distribution example, the mean is α/β and the variance is α/β^2. Thus, we find $\beta = $ mean/variance, and then we can substitute in the value of β to find $\alpha = \beta*$mean. If the standard deviation is given instead of the variance, we can square the standard deviation and proceed as above.

The beta distribution is somewhat more complicated algebraically. The mean is equal to $\alpha/(\alpha + \beta)$ and the variance is a complicated expression in terms of α and β. The expression for the variance can be rewritten more conveniently in terms of the mean as mean$(1-$mean$)/(\alpha + \beta + 1)$, and one can be solve for α and β.

Developing a Nonconjugate (Lognormal) Prior—One of the things that makes the lognormal distribution attractive as a prior in PRA is the ease with which it can encode uncertainty about a parameter that varies over several orders of magnitude. The uncertainty encoded by the lognormal distribution is not usually provided in terms of the distribution parameters (μ and τ) needed by OpenBUGS. More commonly, information is given in terms of a median or mean value and an error factor, or sometimes in terms of an upper and lower bound. Using the properties of the lognormal distribution, any of these sets of information can be translated into the μ and τ parameters needed by OpenBUGS, as shown in the script excerpts below.

```
# Use the following lines if median and error factor given
mu <-log(median)
tau <-pow(log(EF)/1.645, -2)
```

```
# Use the following lines if mean and error factor given
mu <-log(mean) - pow(log(EF)/1.645, 2)/2
tau <-pow(log(EF)/1.645, -2)
```

```
# Use the following lines if median and upper bound given
mu <-log(median)
tau <-pow(log(upper/median)/1.645, -2)
```

```
# Use the following lines if mean and upper bound given
# Caution: mean/upper must be > 0.258
tau <-pow(sigma, -2)
sigma <-(2*1.645 + sqrt(4*pow(1.645, 2) + 8*log(mean/upper)))/2
mu <-log(mean) − pow(sigma, 2)/2
```

```
# Use the following lines if upper and lower bound given
mu <-log(sqrt(upper*lower))
tau <-pow(log(sqrt(upper/lower)/1.645, -2)
```

3.5.2 Developing a Prior from Limited Information

In some cases, not enough information may be available to completely specify an informative prior distribution, as two pieces of information are typically needed. For example, in estimating a failure rate, perhaps only a single estimate is available. This section describes how to use such limited information to develop a distribution that encodes the available information with as much epistemic uncertainty as possible, thus reflecting the limited information available.

As expected, because the information on which the prior is based is very limited, the resulting prior distribution will be diffuse, encoding significant epistemic uncertainty regarding the parameter value. However, it will not be as diffuse as the noninformative priors discussed earlier. Table 3.8 summarizes the results for commonly encountered cases. All of the distributions in this table are conjugate except for the last one.

3.5.3 Cautions in Developing an Informative Prior

We provide some general cautions to consider in developing an informative prior distribution. These and several others are discussed in more detail in [1].

1. *Beware of zero values.*

From Bayes' Theorem, the posterior distribution is proportional to the product of the prior distribution and the likelihood function. Therefore, any region of the parameter space that has zero (or very low) prior probability will also have zero (or very low) posterior probability, regardless of the information supplied by the observed data.

2. *Beware of cognitive biases.*

Because of such biases, distributions elicited from experts can be overly narrow, under-representing uncertainty (this is in some sense a special case of the first caution about avoiding zero values). Beyond this problem, distributions

Table 3.8 Prior distributions encoding limited information about parameter values

Information available	Suggested prior distribution
Mean value for λ in poisson distribution	Gamma distribution with $\alpha = 0.5$ and $\beta = 1/(2 \times \text{mean})$
Mean value for p in binomial distribution	Beta distribution with $\alpha \approx 0.5$ and $\beta = (1 - \text{mean})/(2 \times \text{mean})$
Mean value for λ in exponential distribution	Gamma distribution with $\alpha = 1$ and $\beta = 1/\text{mean}$
p in binomial distribution lies between a and b	Uniform distribution between a and b

elicited from experts can also exhibit systematic bias, which analysts may wish to compensate for.

3. *Ensure that relevant evidence is used to generate the prior.*

This may sound obvious, but it is a common pitfall in our experience. As an obvious example, information pertaining to normal system operation may not be directly relevant to system performance under the more severe operational loads experienced during a PRA accident scenario.

4. *Use care in developing a prior for an unobservable parameter.*

The parameters of the aleatory models are not typically observable. It may be beneficial to develop information for related parameters, such as expected time between events instead of event occurrence rate. Also note that the mean value is a mathematically defined quantity, which may not be a representative value in the case of highly skewed distributions. In such cases, the analyst may wish to use the median instead of the mean in developing a prior distribution.

3.5.4 Ensure Prior is Consistent with Expected Data: Preposterior Analysis

For the final step in selecting an informative prior distribution, we can use the *candidate* prior distribution to generate potential data. We then check to see if it is extremely unlikely that the prior distribution can produce data that are expected (or that have been gathered). If this is the case, then the prior does not encode the analyst's state of knowledge accurately. In the past, such a check was difficult because it requires numerical calculations. However, modern tools such as OpenBUGS make the analysis straightforward. We illustrate the method by way of example.

Assume we are developing a prior for a relief valve's failure-to-open probability. We assume that the prior is a beta distribution with $\alpha = 1.24$ and $\beta = 189,075$. Further, assume that the analyst believes that seeing 2, 3, or possibly even 4 failures over 350 demands would not be surprising. The check is to see if this

Table 3.9 OpenBUGS script for checking a prior distribution via preposterior analysis

```
model {
x ~ dbin(p, n)                          # Binomial likelihood function
p ~ dbeta(alpha, beta)                  # Beta prior distribution for p
p.value <-step(x - x.exp)               # If mean value of this node is < 0.05 or > 0.95, prior is
}                                       # incompatible with expected data
data
list(alpha = 1.24, beta = 189075, x.exp = 2, n = 350)
```

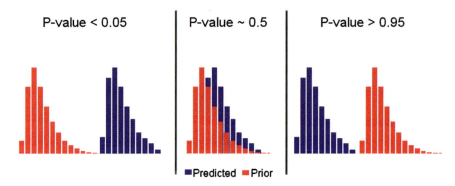

Fig. 3.8 Depiction of the Bayesian p.value calculation used to test prior distribution validity by comparison with a predicted distribution

belief is consistent with the stated beta prior distribution. To perform this check, we need to calculate the probability of seeing 2 or more failures to open in 350 demands on the relief valve (since we expect to see at least 2 failures).

The probability of seeing x failures in n demands is given by the binomial distribution, but this probability is conditional upon a value of p, and p is uncertain. The specified beta prior distribution is a candidate for describing the analyst's epistemic uncertainty about p. If it is compatible with the belief that 2 or more failures in 350 demands are likely, then a weighted-average probability of seeing at least 2 failures in 350 demands, with the weights supplied by the candidate prior distribution, should not be too small.

We use the OpenBUGS script shown in Table 3.9 to estimate the unconditional probability of seeing 2 or more failures in 350 demands, based on the candidate beta prior distribution. Note that this script tests the prior [beta(1.24, 189,075)] since we are not assigning any data to node x.

Running this script for 100,000 iterations with 1,000 burn-in iterations for convergence gives a mean for the p.value node of 0.0 (for two significant digits), indicating that the expected data are *very incompatible* with the candidate beta prior distribution.

This concept of a Bayesian "p-value" is shown in Fig. 3.8, where results with a low (less than 0.05) p-value represent cases where the prior produces a failure count that is much lower than the expected–or predicted–distribution. The converse, when the p-value is high (greater than 0.95), represents cases where the prior produces more failures than expected. The ideal case is obtained when the p-value is close to 0.5—this case represents the prior being close to the expected distribution. Note that the "expected results" could be a constant instead of a distribution and we could check against the posterior (instead of the posterior).

3.6 Exercises

1. A Suppose that X is Poisson(λt), with $\lambda = 0.2$/day and $t = 2$ days. Calculate the probability distribution function of X, at the values $x = 0, 1, 2$, and 3. Then calculate the cumulative distribution function of X.

2. A lognormal distribution with mean 1E-4/year and error factor 5 is going to be used as a prior distribution for a failure rate λ. Data is provided for failures, where two failures were recorded over the last two years of cumulative operation.

    a. Find the 5th and 95th percentiles of the lognormal prior distribution.
    b. Find the mean, 5th and 95th percentiles of the posterior distribution.

3. The following times (in minutes) to repair a component have been collected: 12, 12.7, 16, 20, 25. The prior distribution is a gamma distribution with first parameter equal to 0.5 and a mean value of 0.052/min. Find a posterior mean repair rate and 90% interval.

4. Assume that failures to start of a component can be described by a binomial distribution with probability of failure on demand, p. Using the Jeffreys noninformative prior, and with having observed 3 failures in 400 demands, find the posterior mean and 90% interval of p.

5. Assume there have been 18 failures of pumps to start in 450 demands. For the prior distribution, assume that pump reliability on demand is believed to be no less than 0.95. Find the posterior mean of the failure-to-start probability p, and a 90% credible interval. Assume a binomial model for the number of failures to start in 450 demands.

6. Suppose that the failure rate, λ, for a component is such that $P(\lambda < 10^{-4}$/h$) = 0.05$ and $P(\lambda > 10^{-3}$/h$) = 0.05$. Find the gamma distribution that encodes this prior information about λ.

7. Assume that 32 valve failures were reported in a particular year for 8 facilities. Each facility has 210 such valves and is in operation the entire year.

    a. What is the simplest aleatory model that could be used to described the number of valve failures? What is the unknown parameter in this model?
    b. Find the posterior mean and 90% credible interval for the unknown parameter under the following prior distributions:

    i.   Uniform(0, 10^{-5}/h)

    ii. Jeffreys prior

    iii. Gamma(2.5, 183,200 h)

    iv. Lognormal prior with mean of 13.4 failures/10^6 h and variance $73/10^{12}\,h^2$.

8. Six failures of a certain type of instrument have been observed in 22,425,600 unit-hour of testing. From past experience, the analyst believes the prior probability is 0.05 that the failure rate is less than 10^{-7}/unit-hour, and 0.05 that it exceeds 10^{-5}/unit-hour.

    a. Find the parameters of the gamma distribution that encode this prior information.

    b. Find the posterior distribution for the failure rate.

    c. Find the 90% credible interval for the instrument reliability over a period of 20 yrs.

Reference

1. Siu NO, Kelly DL (1998) Bayesian parameter estimation in probabilistic risk assessment. Reliab Eng Syst Saf 62:89–116

Chapter 4
Bayesian Model Checking

In this chapter, we examine the predictions of our Bayesian inference model ("model" for short) as a test of how reasonable the model is. Recall that the Bayesian inference model comprises the likelihood function (representing aleatory uncertainty AKA our probabilistic model of the world), and the prior distribution (typically representing epistemic uncertainty in parameters in the aleatory model). We begin with direct inference using the posterior distribution, including a brief introduction to Bayesian hypothesis testing. Following this, we will examine how well our model can replicate the observed data; models for which the observed data are highly unlikely to be replicated are problematic and will lead us to alternative prior distributions or likelihood functions, such that the resulting model is better able to replicate the observed data. We adopt the perspective of Gelman et al. [1], which holds that we are not attempting to select the "correct" model. Instead, as George Box famously stated, "All models are wrong, but some are useful." Thus, we are looking for where there are obvious failings in our model that limit its usefulness. In Chap. 8 we present an introduction to model selection based on penalized likelihood criteria, such as the deviance information criterion.

4.1 Direct Inference Using the Posterior Distribution

Consider first an example in which the performance of two components is being compared. This might be done in a variety of contexts. Two common ones would be to question whether data from the two components can reasonably be pooled, or to ask whether one component's performance is significantly better or worse than the other, which is really just another way of phrasing the question about whether the data can be pooled.

Assume as a first example that Component 1 has experienced 3 failures in 75 demands to function, and that Component 2 has experienced 5 failures in 69

D. Kelly and C. Smith, *Bayesian Inference for Probabilistic Risk Assessment*,
Springer Series in Reliability Engineering, DOI: 10.1007/978-1-84996-187-5_4,
© Springer-Verlag London Limited 2011

Fig. 4.1 DAG for comparing performance of two components

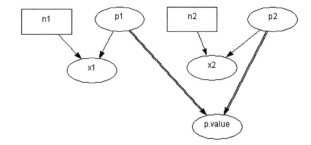

Table 4.1 OpenBUGS script to compare performance of two components

```
model{
  p1 ~ dbeta(0.5,0.5) #Jeffreys prior for p1
  x1 ~ dbin(p1,n1) #Binomial distribution for x1
  p2 ~ dbeta(0.5,0.5) #Jeffreys prior for p2
  x2 ~ dbin(p2,n2) #Binomial distribution for x2
  p.value <- step(p2 - p1) #Gives Pr(p2 > p1)
}
data
list(x1=3, n1=75, x2=5, n2=69)
```

demands. To answer either of the above questions, a Bayesian would calculate $P(p_2 > p_1|\text{data})$, where p_i is the demand failure probability of Component i, $i = 1$, 2. To calculate this probability, one needs the posterior distributions for p_1 and p_2. We will assume the observed failures are described by a binomial distribution in both cases, and we will use the Jeffreys prior for p to focus attention on observed component performance. The DAG for this model is shown in Fig. 4.1, and the OpenBUGS script that implements the model is listed in Table 4.1.

Running this script, we find $P(p_2 > p_1) = 0.80$. This is relatively weak evidence of worse performance of Component 2.

4.2 Posterior Predictive Distribution

Our primary tool for model checking will be the posterior predictive distribution. At times we will use this distribution directly, and at other times we will use summary statistics derived from it. The posterior predictive distribution is the predictive distribution for future values of an observed random variable, given past empirical data. It is defined as:

$$\pi(x_{\text{pred}}|x_{\text{obs}}) = \int_{\Theta} f(x_{\text{pred}}|\theta)\pi_1(\theta|x_{\text{obs}})d\theta \qquad (4.1)$$

In this equation, θ is the parameter of the aleatory model that generates the observed data, x_{obs}. In words, we average the likelihood function for the predicted

Fig. 4.2 DAG representing
posterior predictive
distribution

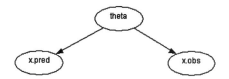

data values over the posterior distribution for θ, to obtain the posterior distribution
for the predicted data, given the observed values. In terms of a DAG model, we
can represent this as shown in Fig. 4.2.

In Fig. 4.2, we observe data, x_{obs}, which updates our prior distribution for θ, the
parameter of the aleatory model that generates x_{obs}. The updated (i.e., posterior)
distribution of θ is then used in the aleatory model to generate predicted data, x_{pred},
whose distribution is given by Eq. 4.1.

We first illustrate this use of the posterior predictive distribution with an
example related to the frequency of an initiating event. Recall that the usual
aleatory model for the occurrence of initiating events is a Poisson distribution with
mean λt, where λ is the initiating event frequency. Thus, the probability of seeing
a specific outcome, such as 1 event in 10 years, or at least 2 events in 25 years, is
obtained from the Poisson distribution:

$$P(X = x|\lambda) = \frac{(\lambda t)^x e^{-\lambda t}}{x!}, \quad x = 0, 1, \ldots$$

If we choose a conjugate gamma prior distribution for the unknown parameter,
λ, the posterior predictive distribution for X_{pred} can be written in closed form, and
evaluated in a spreadsheet. The result can be shown to be:

$$\Pr(X_{pred} = x_{pred}) = \frac{\Gamma(\alpha_1 + x_{pred})}{x_{pred}!\Gamma(\alpha_1)} \left(\frac{t_{pred}}{\beta_1}\right)^x \left(1 + \frac{t_{pred}}{\beta_1}\right)^{-(\alpha_1 + x_{pred})} \tag{4.2}$$

In Eq. 4.2, α_1 and β_1 are the posterior gamma distribution parameters, given in
terms of the prior distribution parameters α_0 and β_0 by $\alpha_1 = \alpha_0 + x$ and
$\beta_1 = \beta_0 + t$.

Let us consider a specific example, taken from Kelly and Smith [2]. Assume the
prior distribution for the initiating event frequency is gamma (1.2, 12 year) and
that the observed data are 3 events in 2.9 years. Updating the prior distribution
with the observed data gives $\alpha_1 = 4.2$ and $\beta_1 = 14.9$ year as the parameters of the
gamma posterior distribution for λ. To check the appropriateness of the model
(specified gamma prior distribution and Poisson likelihood function), we calculate
the probability of observing *at least* 3 events in the next 2.9 years, using the
posterior predictive distribution. The OpenBUGS script in Table 4.2 can be used
to calculate this probability. The probability of 3 or more events is found to be
0.06, which is small enough to suggest there might be problems with the model.

One can also use this technique to help with selecting a prior distribution.
In this case, it is the prior predictive distribution that is employed. The prior
predictive distribution is simply the denominator of Bayes' Theorem. The DAG in

Table 4.2 OpenBUGS script to generate posterior predictive distribution for initiating event example

```
model {
x.obs ~ dpois(mean) #Poisson aleatory model
mean <- lambda*time
lambda ~ dgamma(alpha, beta) #Prior distribution
x.pred ~ dpois(mean) #Posterior predictive node
tail.prob <- step(x.pred - x.obs) #Posterior predictive probability (monitor mean)
}
data
list(x.obs=3, time=2.9, alpha=1.2, beta=12)
```

Table 4.3 OpenBUGS script to calculate prior predictive distribution for valve failure example

```
model {
x.obs ~ dbin(p, n) #Binomial aleatory model
p ~ dbeta(alpha, beta) #Prior distribution
tail.prob <- step(x.obs - x.expected) #Prior predictive probability (monitor mean)
}
data
list(x.expected=2, n=250, alpha=1.2, beta=2.E+5)
```

this case is particularly simple, consisting of a stochastic node for the aleatory parameter connected to a stochastic node representing the data generated by the aleatory model. No data are observed; data expected under the proposed prior distribution and aleatory model are generated from the prior predictive distribution. If the probability calculated for expected data is too small, this suggests an inconsistency between the prior distribution and the expected data. Such analysis is sometimes referred to as preposterior analysis, and was discussed in Chap. 3.

As an example, consider failure of a valve to change state on demand, with a binomial distribution as the aleatory model. Assume the prior distribution under consideration is a beta distribution with $\alpha = 1.2$ and $\beta = 200{,}000$. We judge that 2 failures in 250 demands is a reasonable failure count and want to see if the candidate prior distribution is consistent with this expected data. Again, because the binomial aleatory model and beta prior distribution are conjugate, the prior predictive distribution can be written in closed form. It can be shown that this distribution is given by

$$\Pr(X_{obs} = x_{obs}) = \binom{n}{x_{obs}} \frac{\Gamma(\alpha + x_{obs})}{\Gamma(\alpha)} \frac{\Gamma(\beta + n - x_{obs})}{\Gamma(\beta)} \frac{\Gamma(\alpha + \beta)}{\Gamma(\alpha + \beta + n)} \tag{4.3}$$

Probabilities from Eq. 4.3 can be calculated with a spreadsheet. However, we will use OpenBUGS, as it can also handle nonconjugate priors, where the prior predictive distribution cannot be written down in closed form. Recall that doing the integration in the denominator of Bayes' Theorem (i.e., finding the prior predictive distribution) is the hurdle that has been overcome by modern computational ability, which we exploit fully. The OpenBUGS script is listed in Table 4.3. Running this script in the

Source	Failures	Exposure time year
1	2	15.986
2	1	16.878
3	1	18.146
4	1	18.636
5	2	18.792
6	0	18.976
7	12	18.522
8	5	19.04
9	0	18.784
10	3	18.868
11	0	19.232

Table 4.4 Component data from 11 sources for estimating failure rate

usual way gives a probability of about 0.002 of seeing 2 or more failures in 250 demands. Thus, our candidate beta prior distribution is unlikely to produce the expected data. The 97.5th percentile of x_{obs} with this prior distribution is 0.

4.2.1 Graphical Checks Based on the Posterior Predictive Distribution

Consider the component failure data shown in Table 4.4. We wish to examine whether it is reasonable to pool the data from these 11 sources to estimate a single failure rate, λ, in a Poisson aleatory model. One check we can do is to generate replicate failure counts from the posterior predictive distribution. OpenBUGS can generate a plot of the 95% credible interval from this distribution, with the observed data overlaid. This is obtained from the menu sequence *Inference* $\rightarrow$ *Comparisons* $\rightarrow$ *Model Fit*. This can quickly indicate areas where the model significantly under- or over-predicts the observed variability. For example, by showing where the observed data fall outside the predictive intervals of a model with constant λ, the analyst can diagnose under-prediction of observed variability, which is an indicator of heterogeneous data. Figure 4.3 shows the plot for the data in Table 4.4, based on a Poisson model with constant λ and the Jeffreys prior for λ. This figure clearly indicates that a model with a single λ cannot reproduce the variability in the observed data.

This graphical check can also be used with random durations, such as operator response times, times to recover offsite power, times to repair failed equipment, or times to suppress a fire. Chapter 3 illustrated Bayesian inference for the exponential model for random durations. This is the simplest model for durations, having only one parameter, λ. Chapter 8 will examine more complex models for durations, involving an additional parameter, which allows λ to vary. We will illustrate model-checking with an exponential aleatory model here, with checks for more complicated models covered in Chap. 8. Kelly [3] presents an analysis of the following recovery times (in hours) for a grid-related loss of offsite power at a facility: 0.1, 0.133, 0.183, 0.25, 0.3, 0.333, 0.333, 0.55, 0.667, 0.917, 1.5, 1.517, 2.083, 6.467.

Fig. 4.3 Plot of 95% mean and 95% interval from posterior predictive distribution, illustrating inability of simple Poisson model to replicate variability in data shown in Table 4.4

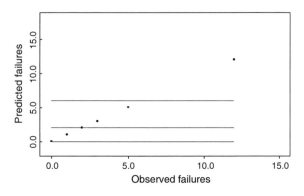

Fig. 4.4 Plot of 95% interval from posterior predictive distribution for exponential model, illustrating inability of model to replicate longest offsite power recovery times from Kelly [3]

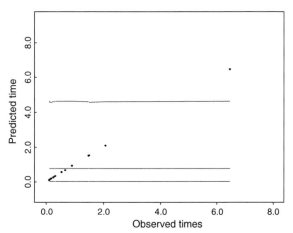

If we assume an exponential aleatory model for these times, with the Jeffreys prior for λ (this focuses attention on the likelihood function), we find the posterior mean of λ to be 0.91/h, with a 90% credible interval of (0.55, 1.35). Figure 4.4 shows the graphical posterior predictive check for these times produced by OpenBUGS. As the figure indicates, the exponential model cannot replicate the longest recovery time, suggesting that a more complex model, which allows a time-dependent recovery rate, may be needed.

4.3 Model Checking with Summary Statistics from the Posterior Predictive Distribution

The frequentist approach to model checking typically involves comparing the observed value of a test statistic to percentiles of the (often approximate) sampling distribution for that statistic. Given that the null hypothesis is true, we would not expect to see "extreme" values of the test statistic. The simplest Bayesian approach to model-checking involves calculating the posterior probability of the various hypotheses and choosing the one that is most likely. However, it is sometimes helpful to use summary statistics[1] derived from the posterior distribution, as described by Gelman et al. [4]. We will discuss two such statistics: a Bayesian chi-square statistic for count data and a Cramer-von Mises statistic for durations. These statistics will lead to a Bayesian analog of the frequentist p-value.

4.3.1 Bayesian Chi-Square Statistic

In frequentist statistics, a commonly encountered test statistic for count data is

$$\chi^2 = \sum_i \frac{(x_i - \mu_i)^2}{\sigma_i^2}$$

In this equation, x_i is the i-th observed value, μ_i is the i-th expected, or mean value, and σ_i^2 is the i-th variance. The distribution of χ^2 is often approximately chi-square, with degrees of freedom related to the sample size. We use this as motivation for the following summary statistics, in which we denote the mean of X by μ.

We use the observed values of X to form the statistic

$$\chi_{\text{obs}}^2 = \sum_i \frac{\left(x_{\text{obs},i} - \mu_i\right)^2}{\sigma_i^2}$$

We then generate predicted values of X from its posterior predictive distribution, and construct an analogous statistic:

$$\chi_{\text{rep}}^2 = \sum_i \frac{\left(x_{\text{pred},i} - \mu_i\right)^2}{\sigma_i^2}$$

Both of these statistics, defined analogously to the frequentist chi-square statistic, but without the need to bin the data, have a posterior distribution. χ_{obs}^2 plays the role of the theoretical distribution in the frequentist setting, and χ_{rep}^2 plays that of the summary statistic based on the data; in this case the "data" are predicted values from the Bayesian model (prior plus likelihood). In the frequentist setting, if the summary statistic calculated from the data is in the tail of the theoretical distribution, we are led to reject our model. The p-value is sometimes used to measure the degree to which the data are in conflict with the model. We will adopt that term here, and define the

---

[1] Strictly speaking, these should not be referred to as *statistics*, as a statistic should not depend upon unknown model parameters. Gelman et al. [4] suggest the term *discrepancy variable*. However, to avoid introducing more terminology than is necessary, we will use *statistic*.

Table 4.5 OpenBUGS script for checking ability of model with single λ to replicate data in Table 4.4

```
model    {
     for (i in 1 : N) {
     lambda[i] <- lambda.const #Constant failure rate
     mean[i] <- lambda[i] * time[i] #Poisson parameter for each source
     x[i] ~ dpois(mean[i]) #Poisson likelihood for failures in each source
     x.pred[i] ~ dpois(mean[i]) #Posterior predictive distribution
     diff.obs[i] <- pow(x[i] - mean[i], 2)/mean[i]
     diff.rep[i] <- pow(x.pred[i] - mean[i], 2)/mean[i]
     }
#Bayesian p-value calculation
chisq.obs <- sum(diff.obs[])
chisq.pred <- sum(diff.pred[])
p.value <-step(chisq.pred - chisq.obs) #Monitor mean value, should be near 0.5
#Jeffreys prior for lambda.const
lambda.const ~ dgamma(0.5, 0.0001)
}
```

Bayesian p-value to be $\Pr(\chi^2_{rep} \geq \chi^2_{obs})$. However, instead of choosing an arbitrary p-value (e.g., 0.05) and rejecting a model with a p-value below this arbitrary cutoff, we will use the p-value to select the model that is best at replicating the observed data. This will be the model with Bayesian p-value closest to 0.5, which is the value one would obtain if the distributions of the observed and replicated test statistics overlapped perfectly.

Consider the 11 sources of data for estimating a failure rate, λ, shown in Table 4.4. Earlier we did a graphical check of whether it is reasonable to pool the data from these 11 sources by graphically checking how well a model with a single λ can replicate the observed data. We will now perform a quantitative check of the model's ability to replicate the observed data.

The OpenBUGS script for checking the model with a single value of λ is listed in Table 4.5. We need to monitor the *p.value* node in this script. If the model with a single λ is adequate, we expect the mean value of this node to be near 0.5; values near 0 or 1 are usually indicative of a problem with the model. We will use the Jeffreys prior for λ to focus attention on the Poisson aleatory model, so problems indicated by the Bayesian p-value may suggest a more complicated aleatory model is required. We will examine more complex alternatives to the simple Poisson model in later chapters.

Running this script in the usual way, we find the mean of the *p.value* node to be 2.3E-4, a very small value, indicating poor ability of a model with a single λ to replicate the data in Table 4.4. This is a strong indicator that a more complex model is needed. The simple Poisson model with a single λ cannot reproduce the variability observed in *x*, as was suggested by the plot shown in Fig. 4.4. The 90% credible interval for the replicated failure count is (0, 6), while the observed failure count ranged from 0 to 12.

Table 4.6 OpenBUGS script implementing Cramer-von Mises statistic for recovery times from [3]

```
model    {
    for (i in 1 : N) {
    time[i] ~ dexp(lambda) #Exponential aleatory model
#Replicate times from posterior predictive distribution
    time.rep[i] ~ dexp(lambda)
    #Rank replicate times
    time.rep.ranked[i] <- ranked(time.rep[], i)
#Calculate components of Cramer-von Mises statistic for observed and replicate data
    F.obs[i] <- cumulative(time[i], time.ranked[i])
    F.rep[i] <- cumulative(time.rep[i], time.rep.ranked[i])
    diff.obs[i] <- pow(F.obs[i] - (2*i-1)/(2*N), 2)
    diff.rep[i] <- pow(F.rep[i] - (2*i-1)/(2*N), 2)
}
#Calculate distribution of Cramer-von Mises statistic for observed and replicate data
CVM.obs <- sum(diff.obs[])
CVM.rep <- sum(diff.rep[])
p.value <- step(CVM.rep - CVM.obs) #Mean value should be near 0.5
#Diffuse prior distributions
lambda ~ dgamma(0.0001, 0.0001)
}
data
list(time=c(0.1,0.133,0.183,0.25,0.3,0.333,0.333,0.55,0.667,0.917,1.5,1.517,2.083,6.467),N=14)
```

4.3.2 Cramer-von Mises Statistic

If the observed variable is a random duration, as in the earlier example of recovery times from Kelly [3], we will use a different type of summary statistic, more appropriate for continuous data. In frequentist inference, a variety of such statistics have been proposed. We will focus on the Cramer-von Mises statistic (Bain and Engelhardt [5]. This statistic compares the empirical cumulative distribution function for the ranked observed and replicated durations, and calculates an overlap probability, analogous to what we did with the chi-square statistic above. The OpenBUGS script in Table 4.6 implements this for the durations listed earlier, which were taken from Kelly [3].

Running this script gives a Bayesian p-value of 0.42, which is quite close to 0.5. This is compatible with Fig. 4.4, where all but the longest of the observed recovery times were well within the 95% credible interval for the replicated times. Kelly [3] examined more complex models, which will be described in Chap. 8, and found that a lognormal model was better able to replicate the observed variability in the recovery times.

4.4 Exercises

1. A licensee is updating the initiating event frequency for loss of turbine-building cooling water. Their prior distribution is lognormal with a mean of 0.02/year and an error factor of 10. They have observed no losses of turbine-building cooling water in 27.5 Rx-years. Carry out the update with a conjugate prior having the same mean and variance as the lognormal prior. Are your results reasonable?

2. A diesel generator is required to have a failure probability (on demand) of 0.05 or less. The facility PRA has estimated the failure probability as being beta-distributed with a mean value of 0.027 and $\beta = 54$. If we observe 3 failures in 27 tests of the EDG, what is the probability that the diesel meets its performance criterion?

3. A plant's PRA is using a distribution for the frequency of loss of main feed-water that is lognormal with a mean of 0.5/Rx-year and an error factor of 3. Over the next 3 years, there are 3 trips of the plant due to loss of main feed-water. The plant's capacity factor during that period is 85%.

   a. What is the probability of seeing 3 or more losses of main feedwater, using the prior mean as a point estimate?
   b. What is the probability of seeing 3 or more losses of main feedwater over this period of time, based on the prior predictive distribution?
   c. What is the probability of seeing 3 or more losses of main feedwater over this period of time, based on the posterior predictive distribution?

4. The following times in minutes are assumed to be a random sample from an exponential distribution: 1.7, 1.8, 1.9, 5.8, 10.0, 11.3, 14.3, 16.6, 19.4, 54.8. Assuming an exponential aleatory model with parameter λ, with the Jeffreys prior for λ, generate replicated times in OpenBUGS and plot the 95% interval for the replicated times overlaid with the observed data. Does this plot show any problems with the assumed exponential model?

5. Use the Cramer-von Mises summary statistic with the above times to compute the Bayesian p-value for an exponential aleatory model with the Jeffreys prior for λ. Does the p-value suggest any problems with the assumed exponential model?

6. Repeat Exercises 4 and 5 for the following sample of times: 0.2, 37.3, 0.5, 4.3, 80.1, 13.3, 2.1, 3.7, 8.2, 2.8.

7. The data below are taken from Siu and Kelly [6], and show failure-on-demand data for emergency diesel generators (EDG) at 10 plants. Perform a check of a binomial aleatory model with constant probability of failure on demand, p, using (a) OpenBUGS caterpillar plot of 95% posterior credible intervals for p, (b) plot of replicated failure counts vs. observed counts, (c) Bayesian p-value based on chi-square summary statistic. For each of these, use the Jeffreys prior for p.

	Failures	Demands
Plant 1	0	140
Plant 2	0	130
Plant 3	0	130
Plant 4	1	130
Plant 5	2	100
Plant 6	3	185
Plant 7	3	175
Plant 8	4	167
Plant 9	5	151
Plant 10	10	150

8. The table below shows successful launches/launch attempts for a series of launch vehicles.

   a. Using a binomial aleatory model for the number of successes for each vehicle, is a common success probability (i.e., no vehicle-to-vehicle variability) a reasonable choice? Use both graphical predictive checks and Bayesian p-value to answer this question.

   b. Let the success probability vary from vehicle to vehicle according to a beta distribution with parameters $K\delta$ and $K(1 - \delta)$, where δ is the expected success probability before any data are observed, and K controls the dispersion of the population variability distribution. Model this in BUGS using a beta (0.5, 0.5) hyper prior for δ and a gamma (5, 1) hyper prior for K. Find the posterior mean and 90% credible interval for success probability of a *future* launch. How does the Bayesian p-value for this model compare with the constant model above? Has the marginal posterior distribution for K been affected much by the observed data? What might this suggest about sensitivity studies for the hyper prior on K?

Vehicle	Outcome
Pegasus	9/10
Percheron	0/1
AMROC	0/1
Conestoga	0/1
Ariane-1	9/11
India SLV-3	3/4
India ASLV	2/4
India PSLV	6/7
Shavit	2/4
Taepodong	0/1
Brazil VLS	0/2

9. Hamada et al. [7] the following projector lamp failure times (in hours) have been collected: 387, 182, 244, 600, 627, 332, 418, 300, 798, 584, 660, 39, 274, 174, 50, 34, 1895, 158, 974, 345, 1755, 1752, 473, 81, 954, 1407, 230, 464, 380, 131, 1205.

   a. The vendor has provided an estimate of the mean time to failure (MTTF) for the lamp. This estimate is 1,000 h. Use this value to develop a prior distribution for the lamp failure rate, assuming the time to failure is exponentially distributed.
   b. Compare the posterior distribution for the failure rate with this prior to what would have been obtained using the Jeffreys noninformative prior for the failure rate.

References

1. Gelman A et al (2004) Bayesian data analysis, 2nd edn. Chapman & Hall/CRC, London/Boca Raton
2. Kelly DL, Smith CL (2009) Bayesian inference in probabilistic risk assessment—the current state of the art. Reliab Eng Syst Saf 94:628–643
3. Kelly DL (2010) Impact of uncertainty on calculations for recovery from loss of offsite power. International Probabilistic Safety Assessment and Management 10. Seattle
4. Gelman A, Meng XL, Stern H (1996) Posterior predictive assessment of model fitness via realized discrepancies. Statistica Sinica 6:733–807
5. Bain L, Engelhardt M (1991) Statistical theory of reliability and life-testing models. Marcel-Dekker, New York
6. Siu NO, Kelly DL (1998) Bayesian parameter estimation in probabilistic risk assessment. Reliab Eng Syst Saf 62:89–116
7. Hamada MS, Wilson AG, Reese CS, Martz HF (2008) Bayesian reliability. Springer, New York

Chapter 5
Time Trends for Binomial and Poisson Data

In this chapter, we will see how to develop models in which p and λ are explicit functions of time, relaxing the assumptions of constant p and constant λ in the binomial and Poisson distribution, respectively. This introduces new unknown parameters of interest into the likelihood function and makes the Bayesian inference significantly more complicated mathematically. However, modern tools such as OpenBUGS make this analysis no less tractable than the single-parameter cases analyzed earlier.

5.1 Time Trend in p

Consider the data shown in Table 5.1, taken from [1]. Following what we have laid out in the earlier chapters as the general approach to Bayesian parameter estimation for PRA, we first carry out a qualitative check to see if there appears to be any systematic time trend in p. To do this, we update the Jeffreys prior for the binomial distribution with the data for each year, and plot the resulting interval estimates of p side by side. This is done via the menu sequence *Inference → Compare → Caterpillar plot*. The Jeffreys noninformative prior is used because we want the resulting intervals to be driven by the observed data; we are focusing attention upon the binomial aleatory model. Note that these are 95% intervals, as this is the coverage produced by OpenBUGS, and cannot be changed easily by the user. The resulting plot is shown in Fig. 5.1, which appears to indicate an increasing trend in p with time, but significant uncertainty in the estimates for each individual year clouds this conclusion. Therefore, additional graphical and quantitative checks are applied, using the techniques described in Chap. 4.

We begin with a plot of the replicated event count for a binomial model with constant p. Again, we use the Jeffreys prior to focus attention on the aleatory

D. Kelly and C. Smith, *Bayesian Inference for Probabilistic Risk Assessment*,
Springer Series in Reliability Engineering, DOI: 10.1007/978-1-84996-187-5_5,
© Springer-Verlag London Limited 2011

Table 5.1 Valve leakage data, taken from [1]

Year	Number of failures	Demands
1	4	52
2	2	52
3	3	52
4	1	52
5	4	52
6	3	52
7	4	52
8	9	52
9	6	52

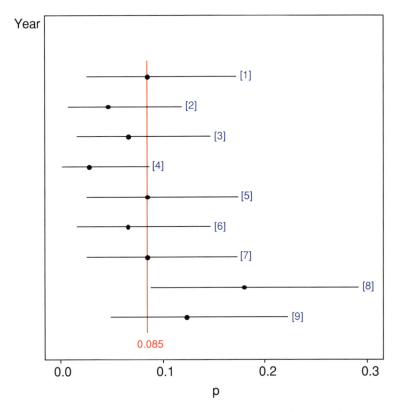

Fig. 5.1 95% posterior credible intervals for valve leakage probability over time, obtained by updating Jeffreys prior for p in each year with the data in Table 5.1. Dots are posterior means for each year, red line is average of posterior means

model. The plot, shown in Fig. 5.2, indicates that this simplest aleatory model has some difficulty replicating the variability in the observed data in Table 5.1. We quantify this using the Bayesian chi-square summary statistic described in Chap. 4. The Bayesian p-value for the binomial distribution with constant p is 0.18, far

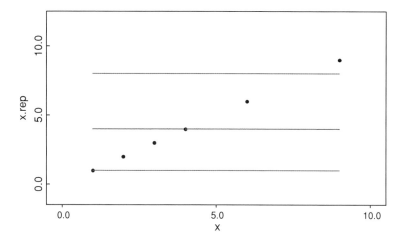

Fig. 5.2 Plot of 95% credible interval for posterior predicted valve leakage events, based on binomial distribution with constant p in each year. Plot indicates difficulty simple model has in replicating the observed data from Table 5.1

enough from the ideal value of 0.5 to suggest that a model should be explored that can incorporate a time trend in p.

A generalized linear model (GLM) is often used to model a monotonic time trend in p or λ. A *link function* is used in such a model to transform the parameter of interest to a related parameter that takes on values over the entire real axis. Various link functions can be used, but a standard choice for p, as suggested by [2], is the logit function. In this model, logit (p), which is the log of the odds ratio, is defined to be a linear function of time:

$$\log\left(\frac{p}{1-p}\right) = a + bt \qquad (5.1)$$

Note that $b = 0$ in this model corresponds to no trend. If p is increasing (decreasing) over time, then we will have $b > (<) 0$. The OpenBUGS script for this model is shown in Table 5.2. To focus attention on the aleatory model, we have used independent improper flat priors over the real axis for a and b; OpenBUGS refers to this distribution as dflat(). Although this is an improper prior, for this type of problem the posterior distribution will be proper and the estimates of a and b will be numerically close to frequentist estimates and the marginal posterior distributions of a and b will be Gaussian. However, in general one should be careful when using the dflat() prior, as it may lead to an improper posterior distribution in some types of problems. This posterior impropriety may be indicated by convergence problems and can often be ameliorated by using a normal distribution with a mean of zero and a very small precision, such as 10^{-6} as an alternative to the improper flat prior.

Because this is a multiparameter model, we will run two chains, starting at dispersed points, as an aid in deciding when convergence to the posterior

Fig. 5.3 Marginal posterior
density for *b*. Values below
zero are very unlikely,
suggesting an increasing
trend in *p* over time

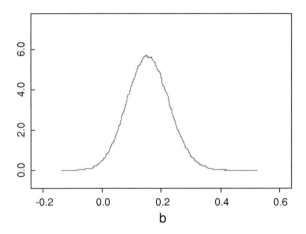

distribution has occurred (see Chap. 6 for more details on checking convergence).
We do this by selecting two chains before pressing the *compile* button during the
model specification step. Also, we must give starting values for each chain, as
OpenBUGS cannot generate initial values of *a* and *b* from the improper flat prior
distributions we have used. These are listed in the *Inits* portion of the script and are
loaded one chain at a time. Frequentist estimates of *a* and *b* can be an aid in
selecting initial values. After loading initial values for the two chains, the *generate
inits* button can be clicked to generate any other initial values required. The model
converges within the first 1,000 iterations.

Now that the two chains have converged, further samples will come from the
desired joint posterior distribution. We run another 100,000 iterations (discarding
the first 1,000 used to achieve convergence) to estimate parameter values.
Examining the posterior density for *b* will help us judge the significance of any
trend that might be present: if the marginal posterior distribution of *b* is mostly to
the right (left) of zero, this indicates an increasing (decreasing) trend. By monitoring the *b* node, we obtain a posterior probability of at least 0.975 that $b > 0$,
suggesting a statistically significant increasing trend in *p*. The plot of the marginal
posterior distribution for *b* in Fig. 5.3 shows this graphically.

We next check the ability of this more complex aleatory model to replicate the
observed data. The posterior predictive plot in Fig. 5.4 indicates that the model with
a logistic time trend in *p* is better able to replicate the observed data than the simple
model without a time trend. The Bayesian *p*-value for this model is 0.47, much closer
to the ideal value of 0.5 than was the value of 0.18 for the model with constant *p*.

Finally, we can use this model to estimate *p* in the next year (year 10). This is
given by node *p*[10] in the script in Table 5.2, and would provide an estimate for
use in a PRA. Monitoring this node, we find a posterior mean of 0.15 and a 90%
interval of (0.085, 0.22). Compare this with the estimates of 0.08 for the mean and
(0.06, 0.099) for the 90% interval from the constant-*p* model, illustrated graphically in Fig. 5.5. Finally, Fig. 5.6 shows the 95% credible intervals for each year,
based on the logistic time-trend model for *p*.

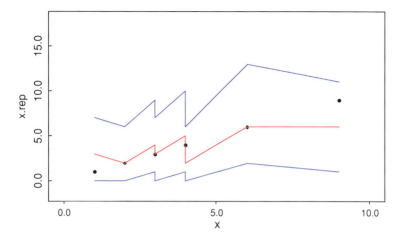

Fig. 5.4 Plot of 95% credible interval for posterior predicted valve leakage events, based on binomial distribution with logistic time trend in p. Observed data are contained within the predictive intervals, suggesting this model is better able to replicate the observed data than the simple model with constant p

Table 5.2 OpenBUGS script for logistic time trend in p

```
model{
    for (i in 1:N) {
      x[i] ~ dbin(p[i], n[i]) #Binomial distribution for failures in each year}
      logit(p[i]) <- a + b*i #Logit link function for p
      #Model validation
      x.rep[i] ~ dbin(p[i], n[i])
      diff.obs[i] <- pow(x[i] - n[i]*p[i], 2)/(n[i]*p[i])
      diff.rep[i] <- pow(x.rep[i] - n[i]*p[i], 2)/(n[i]*p[i])
      }
logit(p[N + 1]) <- a + b*(N + 1) #Used to predict p in 10th year
chisq.obs <- sum(diff.obs[])
chisq.rep <- sum(diff.rep[])
p.value <- step(chisq.rep - chisq.obs)
a ~ dflat() #Laplace priors for a and b
b ~ dflat()
}
```

5.2 Time Trend in λ

As was the case for p in the binomial distribution, it is common to use a GLM to model a monotonic time trend for λ in the Poisson distribution. Here, as discussed by [2], the standard link function is the natural logarithm of λ:

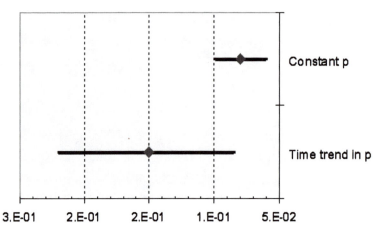

Fig. 5.5 Comparison of time-trend versus constant-probability results for the binomial aleatory model

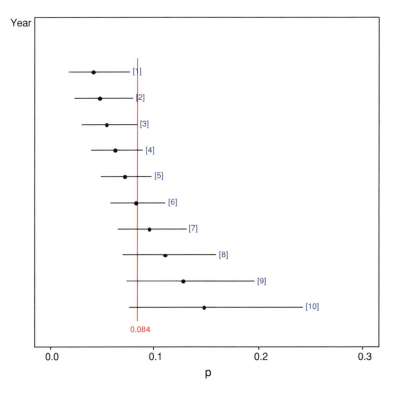

Fig. 5.6 Plot of 95% interval for *p* in each year based on logistic trend model for *p*

Table 5.3 Component
performance data over time,
from [3]

T	Failures	Exp. time
0.5	1	31.64
1.5	1	44.135
3	3	156.407
6	1	608.38
8.5	4	189.545
9.5	9	214.63
10.5	5	216.355
11.5	6	204.091
12.5	12	185.926
13.5	2	157.536
14.5	1	127.608
15.5	3	104.105
16.5	5	77.86
20	3	121.167

$$\log(\lambda) = a + bt \tag{5.2}$$

As was the case for the logistic model for p in the binomial distribution, $b = 0$ in this model corresponds to no trend. If λ is increasing (decreasing) over time, then we will have $b > (<) 0$.

Consider, for example, the component data shown in Table 5.3, taken from [3].

As we did for the first example above, we begin with a simple graphical check, using a caterpillar plot of credible intervals for λ in each time bin, based on the Jeffreys prior for λ in the Poisson distribution. This plot, shown in Fig. 5.7, does not suggest an obvious trend over time. However, it does suggest "random" variability in λ across the time bins.

A posterior predictive plot for a model with constant λ (using the Jeffreys prior) shows some observed data falling outside the 95% intervals of the replicated data, indicating that this simple model may be inadequate. The Bayesian p-value (using the chi-square summary statistic) for this model is 0.004, indicative of an inadequate model (Fig. 5.8).

We now explore a Poisson model with time-dependent λ given by Eq. 5.2. The OpenBUGS script for this model is shown in Table 5.4.

As was the case for the logistic model for p in the binomial distribution, this is a multiparameter model, so we will run two over dispersed chains (see Chap. 6 for details) in order to check for convergence. As was the case in the logistic model for p, convergence occurs within the first 1,000 iterations, and so we run an additional 100,000 iterations for parameter estimation.

Examining the marginal posterior density for b will help us judge the significance of any trend that might be present: if the posterior density of b is mostly to the right of zero, this indicates an increasing trend, and vice versa if the posterior distribution is mostly to the left of zero. The posterior mean of b is 0.097. The 90% credible interval for b is (0.034, 0.16), indicative of an increasing trend for λ over time.

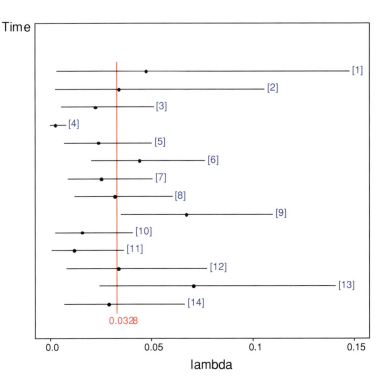

Fig. 5.7 Plot of 95% posterior credible intervals for λ, based on updating Jeffreys prior with data from Table 5.3

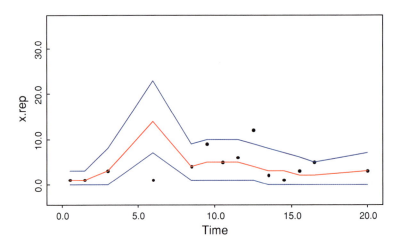

Fig. 5.8 Plot of 95% posterior predictive intervals for replicated data in each time bin. Some observed data (*shown as dots*) fall outside these intervals, suggesting model with constant λ may be inadequate

Table 5.4 OpenBUGS script for Poisson model with loglinear time dependence of λ

```
model {
for (i in 1:N) {
    x[i] ~ dpois(mu[i]) #Poisson dist. for number of failures in each source
#Replicate times from posterior predictive distribution
    x.rep[i] ~ dpois(mu[i]) #Replicate value from post. predictive dist.
    mu[i] <- lambda[i]*ExpTime[i] #Parameter of Poisson distribution
    log(lambda[i]) <- a + b*T[i] #Loglinear link function for lambda
    diff.obs[i] <- pow(F.obs[i] - (2*i-1)/(2*N), 2)
    diff.rep[i] <- pow(F.rep[i] - (2*i-1)/(2*N), 2)
    }
log(lambda[N + 1]) <- a + b*21 #Used to predict lambda in next year
chisq.obs <- sum(diff.obs[])
chisq.rep <- sum(diff.rep[])
p.value <- step(chisq.rep - chisq.obs) #Mean of this node should be near 0.5
p.value <- step(CVM.rep - CVM.obs) #Mean value should be near 0.5
a ~ dflat() #Diffuse priors for a and b
b ~ dflat()
}
```

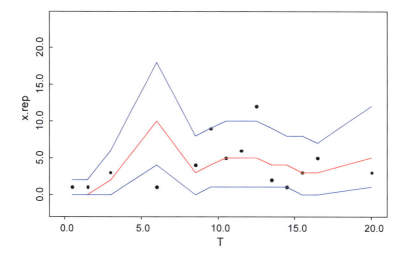

Fig. 5.9 Plot of 95% posterior predictive intervals for replicated data in each time bin. Some observed data (*shown as dots*) fall outside these intervals, suggesting loglinear model for λ may be inadequate

We now carry out posterior predictive checks for this model. The posterior predictive plot shown in Fig. 5.9 is similar to the one for constant λ shown in Fig. 5.8, and suggests a questionable model.

We can also quantify the ability of this model to replicate the observed data by monitoring the mean of the *p*-value node. As before, a mean near 0.5 indicates good replicative ability. The Bayesian p-value for the loglinear model is 0.01. These results are again indicative of a poorly fitting model. Therefore, although we can use this model to predict λ in the next year, we would not want to use this predicted value in a PRA.

This example illustrates the importance of Bayesian model checking. It is easy to fit a time-dependent model for p or λ, and such a model may appear to suggest a significant trend, as was the case in this last example, where there was a high probability that $b > 0$. However, there is no guarantee that such a model has any predictive validity; if it cannot even reproduce the observed data with reasonable probability, as in the last example, then it certainly should not be used to predict performance in the future.

References

1. Pulkkinen U, Simola K (2000) Bayesian models and ageing indicators for analysing random changes in failure occurrence. Reliab Eng Syst Saf 68:255–268
2. Atwood C, LaChance JL, Martz HF, Anderson DJ, Engelhardt ME, Whitehead D et al. (2003) Handbook of parameter estimation for probabilistic risk assessment. U. S. Nuclear Regulatory Commission
3. Rodionov A, Kelly D, Uwe-Klügel J (2009) Guidelines for analysis of data related to ageing of nuclear power plant components and systems. Joint Research Centre, Institute for Energy, Luxembourg: European Commission

Chapter 6
Checking Convergence to Posterior Distribution

OpenBUGS uses Markov chain Monte Carlo (MCMC) sampling to generate values directly from the target posterior distribution. There are theoretical results that guarantee convergence to the posterior distribution, under very general conditions, as described in [1]. However, from a practical perspective it takes time for the MCMC sampling to converge to the posterior distribution; any values sampled prior to convergence should not be used to estimate parameter values. For simple problems involving one parameter, such as p in the binomial distribution or λ in the Poisson distribution, 1,000 iterations will be more than sufficient for convergence. In more complicated problems, which usually involve inference for more than one parameter, this may not be the case, and the user will have to check for convergence. This chapter presents qualitative and quantitative convergence checks that an analyst can use, using features built into OpenBUGS. We address three important issues related to convergence:

1. Convergence to the joint posterior distribution,
2. Coverage of the parameter space,
3. Number of samples needed after convergence.

6.1 Qualitative Convergence Checks

For problems with more than one parameter, which are the problems in which convergence becomes an issue, we recommend that at least two chains be used, with starting values that are dispersed around the estimated mode of the posterior distribution. Usually the analysis will not be very sensitive to the initial values selected for the chains, but this is not always the case. For example, when modeling population variability (see Chap. 7), there are mathematical approaches that can be used to select initial values.

D. Kelly and C. Smith, *Bayesian Inference for Probabilistic Risk Assessment*,
Springer Series in Reliability Engineering, DOI: 10.1007/978-1-84996-187-5_6,
© Springer-Verlag London Limited 2011

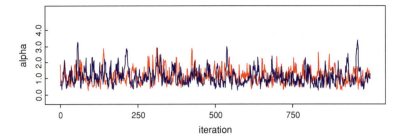

Fig. 6.1 History plot showing two well-mixed chains, indicative of convergence

Fig. 6.2 History plot
showing two poorly-mixed
chains, indicative of failure to
converge

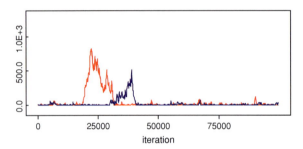

After the model has been compiled (remember to specify the number of chains before compiling the model) and initial values have been loaded, specify the nodes to be monitored. All parameters should be monitored and convergence should be checked for each of these monitored nodes. Now run 100–1,000 samples and select *History* from the *Inference* menu in OpenBUGS to generate a trace of these samples for each monitored node. A plot like the one shown in Fig. 6.1, in which the two chains are well mixed, is indicative of convergence.

In contrast, Fig. 6.2 shows a case in which the chains are not well mixed, which indicates that more iterations must be run to achieve convergence.[1]

6.2 Quantitative Convergence Checks

OpenBUGS has a built-in convergence diagnostic that can be used in conjunction with the history plots shown above to help the user decide when enough burn-in samples have been taken. It requires two or more chains, and is based on an

---

[1] It is possible, especially with highly correlated parameters, that there will be difficulty in getting the chains to mix, despite convergence. Since we cannot readily distinguish between the two problems, we will refer to poor chain mixing as being a sign of failure to achieve convergence. Regardless of the source of the lack of mixing, the estimates should not be used until the problem is rectified, perhaps by reparameterizing the problem in terms of parameters that are less strongly correlated.

Fig. 6.3 BGR plot
illustrating convergence

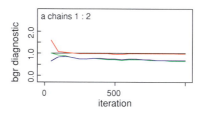

Fig. 6.4 BGR plot of history
in Fig. 6.2, illustrating lack of
convergence

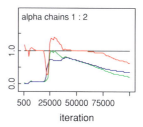

analysis of the variance within- and between-chains. If the chains have converged,
all chains should have approximately the same within-chain variance, and this
should be approximately equal to the between-chain variance estimate. The
Brooks-Gelman-Rubin (BGR) diagnostic in OpenBUGS looks at a ratio of these
estimates, which is normalized to equal one when the chains have converged.

To implement the BGR diagnostic, run at least two chains, as described above,
then select *bgr diag* from the *Inference* menu. The resulting plot will have three
curves. The estimate of the within-chain variance is shown in blue, the between-
chain estimate is in green, and the normalized BGR ratio is shown in red. The
BGR ratio is expected to start out greater than one if the initial values are widely
dispersed. The heuristic is that this ratio should be less than about 1.2 for convergence. In addition, the between-chain and within-chain estimates shown by the
green and blue curves should be stable. Right-clicking on the BGR graph allows
the analyst to bring up a table of the values over the history.

Figure 6.3 shows a typical BGR plot for a problem that has converged.
Figure 6.4 shows the plot of BGR for the history in Fig. 6.2, confirming the failure
to converge suggested by the history plot.

6.3 Ensuring Adequate Coverage of Posterior Distribution

The second issue the analyst faces is whether the chains are providing good
coverage of the posterior distribution. It can happen, especially because the
samples from MCMC are dependent, that a chain becomes stuck in a particular
region of the parameter space. If this happens, the resulting parameter estimates
can be substantially in error. Running multiple chains with widely dispersed initial
values is the first line of defense against this problem. OpenBUGS also calculates a

Fig. 6.5 Plot of lag
autocorrelation coefficient
exhibiting rapid decrease to
zero with increasing lag

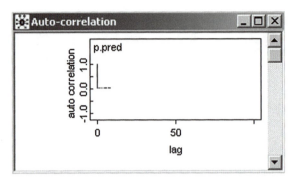

Fig. 6.6 Plot of lag
autocorrelation that may be
indicative of poor coverage of
posterior distribution

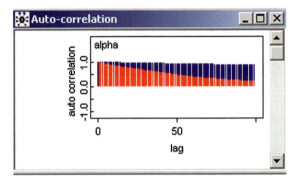

lag autocorrelation coefficient, which measures the degree of dependence between samples from the chains. Ideally, this dependence will fall to zero very quickly as the lag increases; high correlation over long distances in the chain can be an indicator of inadequate coverage of the full posterior distribution. Figure 6.5 shows a plot of the lag autocorrelation that is indicative of good coverage, because the dependence falls quickly to zero as the lag increases. Conversely, Fig. 6.6 shows a plot in which the MCMC samples exhibit a high degree of dependence, even with high lag. Such a plot may be indicative that the chains are not providing good coverage of the posterior distribution.

6.4 Determining Adequate Sample Size

The third issue related to convergence is determining how many samples to take following burn-in (i.e., convergence to the posterior distribution). The discussion in this section is predicated upon having good coverage of the posterior distribution.

The uncertainty of a parameter estimate can be decomposed into two parts, measured by the variance, or its square root, the standard deviation: the "true" uncertainty and the additional uncertainty introduced by Monte Carlo sampling. OpenBUGS computes a sample standard deviation for each monitored stochastic node, which is a measure of the overall uncertainty. It also computes "MC error," which is a measure of the second component, the uncertainty introduced by Monte Carlo sampling. We need to have enough samples after convergence so that MC error is a small contributor to the overall standard deviation, no more than a few percent. The OpenBUGS User Manual suggests 5% as an upper limit; in practice, for the types of problems encountered in PRA, we feel that 2–3% is a better guideline.

Reference

1. Robert CP, Casella G (2010) Monte Carlo statistical methods, 2nd edn. Springer, Berlin

Chapter 7
Hierarchical Bayes Models for Variability

This chapter discusses the Bayesian framework for expanding common likelihood functions introduced in earlier chapters to include additional variability. This variability can be over time, among sources, etc.

7.1 Variability in Trend Models

Chapter 5 introduced problems in which an aleatory model parameter such as p or λ is allowed to vary monotonically over time. Consider the example data in Table 7.1. If we assume a Poisson aleatory model for the number of events in each age period, with parameter $\lambda_i t_i$, and update the Jeffreys prior for λ with the event count for each age period, we get the caterpillar plot shown in Fig. 7.1, which suggests that λ may be decreasing with time.

If we model the apparent time trend in λ using the loglinear model introduced in Chap. 5, with flat prior distributions on the loglinear coefficients a and b, we find that the marginal posterior distribution for the slope parameter, b, lies almost entirely below zero, appearing to indicate a decreasing trend in λ, in agreement with Fig. 7.1. However, if we estimate the Bayesian p-value for such a model as in Chaps. 4 and 5, we find a value of about 0.009, a quite small value, suggesting that the loglinear model for λ is not very good at replicating the observed data. The plot of the predicted event count for each age bin under this model (Fig. 7.2) bears out the low Bayesian p-value. As can be seen in Fig. 7.2, a number of observed event counts are at the limit (especially the lower limit) of the 95% credible interval for the replicated event count, suggesting that the loglinear model has some difficulty replicating the observed data.

In an attempt to explore a richer model that could give rise to the data in Table 7.1, let us consider an extension to the loglinear model, in which there is still a monotonic trend with age, but with additional random variability in λ over time. Our expanded trend model becomes

D. Kelly and C. Smith, *Bayesian Inference for Probabilistic Risk Assessment*,
Springer Series in Reliability Engineering, DOI: 10.1007/978-1-84996-187-5_7,
© Springer-Verlag London Limited 2011

Table 7.1 Example data showing apparent decreasing trend over time in λ

x	Age	Exposure time
19	0.5	0.71
3	1.5	0.75
6	2.5	0.79
0	3.5	0.90
1	4.5	0.82
3	5.5	0.75
0	6.5	1.00
0	7.5	0.83
2	8.5	0.84

Fig. 7.1 Caterpillar plot showing apparent decreasing trend in λ over time

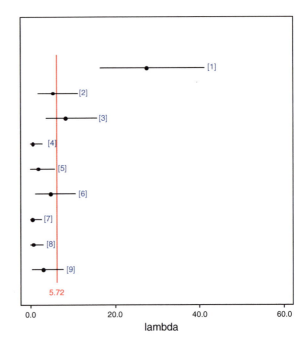

Fig. 7.2 Plot of 95% credible interval (*blue lines*) for predicted event count under loglinear model. Dots are observed event counts

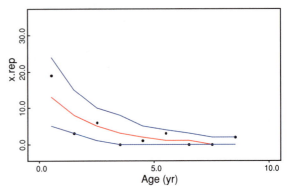

Table 7.2 OpenBUGS script for loglinear model for λ with extra-Poisson variation in each time period

```
model
{
for (i in 1:N) {
          log(lambda[i]) <- a + b*age[i] + eps[i]
          eps[i] ~ dnorm(0, tau.eps)
#Poisson aleatory model and predictive distributions for model checking
          mu[i] <- lambda[i]*time[i]
          x[i] ~ dpois(mu[i])
          x.rep[i] ~ dpois(mu[i])
          diff.obs[i] <- pow(x[i] - mu[i], 2)/mu[i]
          diff.rep[i] <- pow(x.rep[i] - mu[i], 2)/mu[i]
          }
#Model checking
chisq.obs <- sum(diff.obs[])
chisq.rep <- sum(diff.rep[])
p.value <- step(chisq.rep - chisq.obs)
#Prior distributions
a ~ dflat() #Diffuse prior on a
b ~ dflat() #Diffuse prior on b
tau.eps <- pow(sigma.eps, -2)
sigma.eps ~ dunif(0, 10)
}
```

$$\log(\lambda_i) = a + bt_i + \varepsilon_i \qquad (7.1)$$

In this equation, ε_i is a random error term in each age period, which we will take to be normally distributed with mean zero and constant (unknown) variance in each period. We will perform Bayesian inference on the unknown variance, along with the parameters of the loglinear trend model. The OpenBUGS script for this model is shown in Table 7.2. OpenBUGS parameterizes the normal distribution in terms of the mean and precision, which is the reciprocal of the variance. In this script we place the prior on the standard deviation in each time period, which is the square root of the variance. Note the use of a diffuse uniform distribution for the standard deviation. A uniform distribution is used to avoid problems one can encounter using the nearly improper gamma(0.0001, 0.0001) distribution, as discussed in [1].

If we run two MCMC chains, initialized at ($a = 5$, $b = -0.1$) and ($a = -5$, $b = 0.1$), we find that about 2,000 samples are needed to be reasonably sure of convergence (see Chap. 6 for details on checking for convergence). Running an additional 100,000 samples gives adequately low Monte Carlo error for parameter estimation. With variability introduced in each age period, the marginal posterior distribution for b now extends above 0, with a 95th percentile of 0.03. So while there is still evidence for a decreasing trend in λ, it is weaker than in the first case, which omitted extra-Poisson variability. The Bayesian p-value increases to 0.39, indicative of a model that is substantially better at replicating the observed data

Fig. 7.3 Plot of 95%
credible interval (*blue lines*)
for predicted event count
under loglinear model with
additional random variability
in λ. Dots are observed event
counts

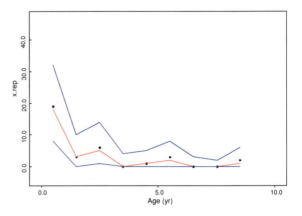

Table 7.3 Component
performance data over time,
from [2]

Time	Failures	Exp. time
0.5	1	31.64
1.5	1	44.135
3	3	156.407
6	1	608.38
8.5	4	189.545
9.5	9	214.63
10.5	5	216.355
11.5	6	204.091
12.5	12	185.926
13.5	2	157.536
14.5	1	127.608
15.5	3	104.105
16.5	5	77.86
20	3	121.167

than the loglinear model with its p-value of 0.009. The plot of credible intervals for
the replicated event count in Fig. 7.3 graphically illustrates the improved predictive ability of this model in comparison with the loglinear model without additional variability in λ over time.

This problem illustrates the so-called *hierarchical Bayes* approach. Hierarchical
Bayes is so-named because it utilizes hierarchical or multistage prior distributions.
In each age bin in this example, the distribution of λ is conditional upon a value
of ε, the error term describing the random year-to-year variation in log(λ) about
a straight line. But ε is uncertain, and we model this uncertainty by introducing
a prior distribution for the standard deviation of ε, which is assumed to be constant
across time. Thus we have a hierarchy of prior distributions.

Recall from Chap. 5 the example using data from [2], reproduced in Table 7.3.
In Chap. 5, we considered two aleatory models for these data: a simple Poisson
model with constant λ, and a time-dependent Poisson model with a loglinear
trend in λ. However, neither of these models performed well on either qualitative

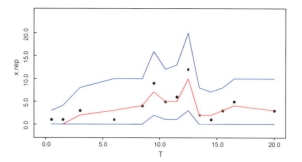

Fig. 7.4 Plot of 95% posterior predictive intervals for replicated data in each time bin. All observed data fall within these intervals, suggesting good predictive ability for a loglinear model with additional variability in λ over time

or quantitative posterior predictive checks. The Bayesian p-values were 0.004 for the constant-λ model, and 0.01 for the loglinear model.

Let us now examine a hierarchical time trend model, as above, where we introduce additional variability in λ over time. Using the script in Table 7.2, we find a less significant trend over time than with the simple loglinear trend model used in Chap. 5. The 90% credible interval for b, which was (0.034, 0.16) for a simple loglinear model, is now (−0.04, 0.14). The predictive validity has improved with the addition of variability over time, as shown by the posterior predictive plot in Fig. 7.4, and the Bayesian p-value has increased to 0.43.

7.2 Source-to-Source Variability

A more common past application of hierarchical Bayes analysis in PRA has been as a model of variability among data sources, for example variability in emergency diesel generator (EDGs) performance across plants, or across time. Such an example of source-to-source variability is discussed at length in [3] from the perspective of two-stage Bayes and empirical Bayes, which are both approximations to a hierarchical Bayes treatment, and have been commonly used in past PRAs, before the availability of tools such as OpenBUGS, which make a fully hierarchical Bayes analysis tractable. The hierarchical Bayes approach to the same problem is presented in [4]. Other examples related to PRA can be found in [4–6].

We will first examine the EDG example from [3]. The data for EDG failures are reproduced in Table 7.4. At each plant, the number of observed failures, X_i, is modeled with a binomial aleatory model with parameters n_i and p_i. To develop a hierarchical model for p, we specify a *first-stage prior*, which is often of a particular functional form, often a conjugate prior, although this is not necessary. For this example, we will take the first-stage prior to be a conjugate beta (α, β) distribution. We now need to specify a prior distribution on the first-stage parameters, α and β. This is called the *second-stage prior*, or *hyperprior*. Note that although nothing limits the analysis to two stages, the use of more than two stages has been rare in PRA applications. It is common, although not necessary, to employ independent, diffuse distributions at the second-stage.

Table 7.4 EDG failure data for 10 plants, taken from [3]

Plant	Failures	Demands
1	0	140
2	0	130
3	0	130
4	1	130
5	2	100
6	3	185
7	3	175
8	4	167
9	5	151
10	10	150

Fig. 7.5 DAG for hierarchical Bayes model of source-to-source variability

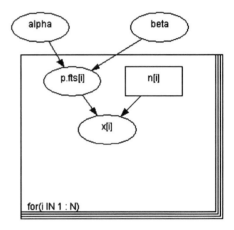

Hierarchical problems such as this one are most simply represented via a DAG. The DAG for this first example is shown in Fig. 7.5. Note in this figure that α and β are independent, prior to the observation of the data. This obviates the need to develop a joint prior distribution for α and β that includes dependence. Once the data are observed, these nodes become dependent, and the joint posterior distribution will reflect this dependence. Note that in some cases the high degree of correlation between α and β can lead to very slow convergence to the joint posterior distribution. In such cases, it may be helpful to reparameterize the problem in terms of the mean and a dispersion measure, which are approximately independent in the joint posterior distribution. We will examine this situation in more detail below.

Another point worth noting about the DAG in Fig. 7.5 is that the EDG failure probabilities (p.fts[i]) are conditionally independent, given values of α and β. The posterior predictive distribution of p, representing source-to-source variability, will be given by an average of the posterior distribution for p, conditional upon α and β (a beta distribution), weighted by the joint posterior distribution for α and β. We can take advantage of the fact that, as Fig. 7.5 illustrates, the components p_i are conditionally independent, given α and β, to write

Table 7.5 OpenBUGS script for modeling plant-to-plant variability in EDG performance

```
model {
for (i in 1 : N) {
          p[i] ~ dbeta(alpha, beta) #First-stage prior
          x[i] ~ dbin(p[i], n[i]) #Binomial dist. for failures at each plant
          }
p.pred ~ dbeta(alpha, beta)
alpha ~ dgamma(0.0001, 0.0001) #Vague hyperprior for alpha
beta ~ dgamma(0.0001, 0.0001) #Vague hyperprior for beta
inits
list(alpha = 1, beta = 25)
list(alpha = 0.5, beta = 75)
}
```

$$\pi\left(p_i | \tilde{x}, \, \tilde{n}\right) = \int_0^1 \int_0^1 \cdots \int_0^1 \left\{ \int \int \left[\prod_{i=1}^N \pi_1\left(p_i | \alpha, \, \beta\right) \right] \pi_2\left(\alpha, \, \beta | \tilde{x}, \, \tilde{n}\right) d\alpha d\beta \right\}$$

$$dp_1 dp_2 \cdots dp_{i-1} dp_{i+1} \cdots dp_n$$

$$= \int \int \pi_1\left(p_i | \tilde{x}, \tilde{n}, \, \alpha, \, \beta\right) \pi_2(\alpha, \, \beta | \tilde{x}, \, \tilde{n}) d\alpha d\beta$$

The second line in this equation is obtained by interchanging the order of integration. Thus, the marginal posterior distribution for p at any particular plant is a continuous mixture of beta distributions, mixing over the joint posterior distribution of the hyperparameters, α and β. The distribution describing plant-to-plant variability in p is the posterior predictive distribution, sometimes referred to in PRA (especially older references) as the average population variability curve:

$$\pi(p^* | \tilde{p}) = \int \int \pi_1\left(p^* | \alpha, \, \beta, \, \tilde{\lambda}, \, \tilde{x}, \, \tilde{n}\right) \pi_2(\alpha, \, \beta | \tilde{x}, \, \tilde{n}) d\alpha d\beta$$

$$= \int \int \pi_1(p^* | \alpha, \, \beta, \, \tilde{x}, \, \tilde{n}) \pi_2(\alpha, \, \beta | \tilde{x}, \, \tilde{n}) d\alpha d\beta$$

It is thus a similar mixture of beta distributions. It is generated in OpenBUGS (node p.pred in the script) by sampling α and β from their joint posterior distribution, and then sampling p^* from the first-stage prior, a beta distribution in this example. The OpenBUGS script used to analyze this problem is shown in Table 7.5.

Two Markov chains, each starting from a separate point in the parameter space, were used with this script. More than two chains may be useful in some problems, although two are sufficient for this example. More than one chain aids in checking convergence, as discussed in Chap. 6. Each of the chains must be given an initial value of α and β. Reference [4] discuss the use of empirical Bayes as an aid in selecting starting values for the chains. The empirical Bayes estimates are 1.2 for α and 63 for β, and the initial values shown in the script in Table 7.5 were dispersed around these values. Note that the empirical Bayes estimates can be calculated using the online calculator at https://nrcoe.inel.gov/radscalc/Default.aspx.

Fig. 7.6 Summary of
marginal posterior
distributions for p at each
of the 10 plants

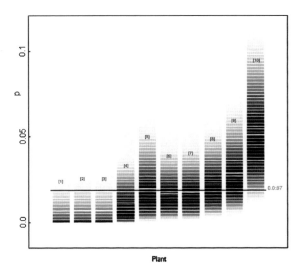

Table 7.6 Results for EDG #1, with comparison of other methods from [3]

	5th	50th	95th	Mean
Empirical Bayes	4.7E-04	4.4E-03	1.7E-02	5.9E-03
Two-stage Bayes	1.2E-04	3.3E-03	1.8E-02	5.2E-03
Hierarchical Bayes	5.9E-05	4.5E-03	1.9E-02	6.3E-03

The model converges quickly: 2,000 iterations are sufficient for burn-in.
Running an additional 100,000 iterations with which to estimate parameter values
gives a posterior predicted mean for p.pred of 0.02, with a 90% credible interval
of (3.7E-4, 0.06). Figure 7.6 summarizes the marginal posterior distribution of
p for each of the 10 plants. The marginal posterior distribution for p for EDG #1 is
summarized in Table 7.6, along with the results of the other approaches described
in [3]. The results from the hierarchical Bayes analysis are generally comparable to
those from empirical and two-stage Bayes, but with somewhat wider uncertainty
bounds. Also, as noted by [3], α and β are highly correlated in the posterior
distribution: the rank correlation coefficient calculated by OpenBUGS is 0.98.
Note that this correlation is automatically accounted for in the MCMC sampling
process in OpenBUGS.

We consider next a similar example for a Poisson aleatory model. Table 7.7
presents component failure data for 11 sources. Figure 7.7 is a caterpillar plot of
95% credible intervals for λ for each of these sources, based on updating the
Jeffreys prior for λ with the data from each source. The lack of overlap of the
intervals in this figure suggests that there may be extra-Poisson variation among
these sources, so that a model with a single λ may not be adequate.

Before we rush to use a hierarchical model in this case, we pause to do
some posterior predictive checking of a simple Poisson aleatory model with
a single λ, using the Jeffreys prior to focus attention on the aleatory model.

Table 7.7 Component
failure rate data for
hierarchical Bayes example

Source	Failures	Exposure time (year)
1	2	15.986
2	1	16.878
3	1	18.146
4	1	18.636
5	2	18.792
6	0	18.976
7	12	18.522
8	5	19.04
9	0	18.784
10	3	18.868
11	0	19.232

Fig. 7.7 Posterior 95%
credible intervals for λ for
each of the 11 sources shown
in Table 7.7, illustrating
apparent extra-Poisson
variability among the sources

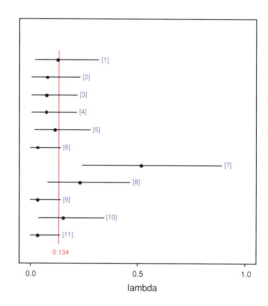

The plot of replicated vs. observed event counts in Fig. 7.8 suggests that
a model with a single value of λ cannot reproduce the variability seen in the
observed data. The Bayesian p-value for this simple model is about 0.0002,
confirming its poor validity.

Because the simple Poisson model has poor predictive validity with respect
to the observed data, we turn to a hierarchical Bayes model. This model is
exactly analogous to that used for the EDGs in the first example. In this
case, because the aleatory model is Poisson instead of binomial, we will use
a gamma(α, β) distribution as the first-stage prior, although a conjugate first-
stage prior is again not necessary. The OpenBUGS script for this model is
shown in Table 7.8.

Fig. 7.8 Plot of 95%
intervals for replicated event
counts, illustrating inability
of model with single value of
λ to replicate observed
variability in the data in
Table 7.7

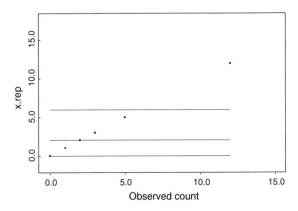

Table 7.8 OpenBUGS script for hierarchical Bayes analysis of variability in λ

```
model {
for (i in 1 : N) {
          lambda[i] ~ dgamma(alpha, beta) #Model variability in LOSP frequency
          mean[i] <- lambda[i] * time[i] #Poisson parameter for each plant
          x[i] ~ dpois(mean[i]) #Poisson dist. for events at each plant
          }
lambda.pred ~ dgamma(alpha, beta)
alpha ~ dgamma(0.0001, 0.0001) #Vague hyperprior for alpha
beta ~ dgamma(0.0001, 0.0001) #Vague hyperprior for beta
inits
list(alpha = 1, beta = 1)
list(alpha = 0.5, beta = 5)
}
```

Two Markov chains, each starting from a separate point in the parameter space,
were used in the analysis, as above. We again used empirical Bayes as an aid in
selecting starting values for the two chains. The empirical Bayes estimates are 0.85
for α and 6.4 year for β.

The model converges quickly: 2,000 iterations are sufficient. Running an
additional 100,000 iterations with which to estimate parameter values gives
a posterior predicted mean for `lambda.pred` of 0.16, with a 90% credible
interval of (9.8E-4, 0.53). Figure 7.9 summarizes the marginal posterior distribution of λ for each of the 11 sources.

The posterior predictive plot shown in Fig. 7.10 illustrates the enhanced ability
of the hierarchical model to replicate the variability in the observed data. This is
reinforced by the Bayesian p-value of 0.45 for the hierarchical model.

Fig. 7.9 Summary of
marginal posterior
distribution for λ for the 11
sources listed in Table 7.7

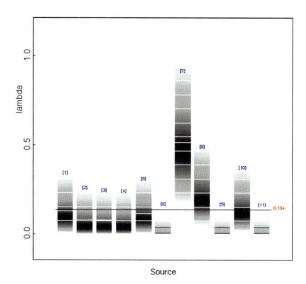

Fig. 7.10 Plot of 95%
intervals for replicated event
counts, illustrating ability of
hierarchical Bayes model of
source-to-source variability
to replicate data in Table 7.7

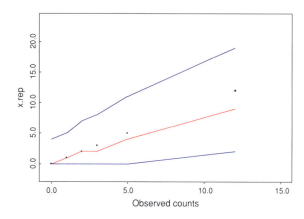

7.3 Dealing with Convergence Problems in Hierarchical Bayes Models of Variability

We turn now to an example in which convergence to the posterior distribution is not so rapid.[1] In our experience, convergence issues with hierarchical Bayes models for population variability make them among the most challenging of common PRA Bayesian inference problems from a computational perspective.

---

[1] As noted earlier, the problem illustrated here is really a problem of poor mixing of the chains due to high correlation between the parameters of the second-stage prior; however, the effect is the same and must be ameliorated before the MCMC samples can be used for parameter estimation.

Fig. 7.11 Side-by-side plot
of 95% credible intervals for
23 data sources in Table 7.9

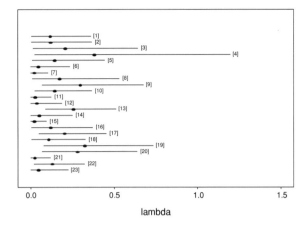

Modern tools such as OpenBUGS have eliminated the computational challenges of the high-dimensional integrals encountered in these kinds of problems, but the burden is still upon the analyst to be on the alert for convergence problems. We use the following example to illustrate the problem and a practical solution.

Consider the 23 sources of data for loss of offsite power, taken from [7]. shown in Table 7.9.

Let us assume that the failure count for each source is Poisson-distributed with rate λ_i. Figure 7.11 shows a side-by-side plot of the 95% credible intervals for λ for each of these sources, based on updating the Jeffreys prior, and illustrates the variability in the sources. However, most of the intervals overlap, suggesting that the source-to-source variability is not large.

Let us first adopt a conjugate first-stage prior, which in this case is a gamma distribution. As above, empirical Bayes can be used as an aid in specifying initial values of α and β. The empirical Bayes estimates are 2.1 and 23 year, respectively. Using these point estimates of α and β in the first-stage gamma prior that describes source-to-source variability, the average rate is 0.09/year, with a 90% credible interval of (0.02, 0.2).

Turning now to hierarchical Bayesian analysis using OpenBUGS with the script given in Table 7.8 and the data in Table 7.9, we must specify hyperpriors for α and β. We will use independent, diffuse distributions, so that the data drive the results. A gamma distribution with both parameters very small gives a distribution that is essentially flat over the positive real axis. With both parameters of the gamma distribution equal to 10^{-4}, the 5th percentile is zero, effectively, and the 95th percentile is approximately 10^{-220}. Thus, there is a very sharp vertical asymptote at zero, and the density is essentially flat (and approximately equal to zero) for values greater than zero.

Following the guidance for checking convergence provided in Chap. 6, we run multiple chains in order to monitor convergence, both graphically via chain history plots, and quantitatively, using the BGR convergence diagnostic calculated by

Table 7.9 Example data from [7]

Events	Exposure time (year)
1	13.054
1	12.77
1	7.22
1	3.944
1	10.548
0	10.704
0	24
1	8.76
3	11.79
2	17.5
0	20.03
0	13.39
5	21.5
0	10.075
0	26.32
1	12.54
3	17.5
1	14.3
3	10.89
3	12.5
0	21.38
2	19.65
0	11.34

Fig. 7.12 History of 100,000 iterations for α, showing failure of chains to mix

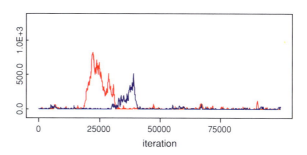

OpenBUGS. OpenBUGS cannot generate initial values from the diffuse gamma hyperpriors used for α and β, so initial values must be supplied. We start with the point estimates from the empirical Bayes analysis above, and pick values around these, as these approximate the mode of the joint posterior distribution for α and β.

Figure 7.12 shows the history of 100,000 samples for α. The two chains are not well mixed, indicating potential convergence problems. The corresponding BGR plot is shown in Fig. 7.13. If the chains have converged to the posterior distribution, the normalized BGR ratio (red line) should be about 1.0, and all three lines

Fig. 7.13 BGR diagnostic
for first 100,000 iterations for
α, illustrating failure to
converge

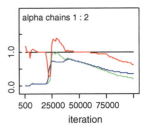

Table 7.10 OpenBUGS script for hierarchical Bayes model for λ, parameterized in terms of mean and coefficient of variation of gamma first-stage prior

```
model {
for (i in 1 : N) {
          lambda[i] ~ dgamma(alpha, beta) #Model variability in LOSP frequency
          mu[i] <- lambda[i] * time[i] #Poisson parameter for each plant
          x[i] ~ dpois(mean[i]) #Poisson dist. for events at each plant
          }
lambda.pred ~ dgamma(alpha, beta)
alpha <- pow(COV, -2)
beta <- alpha/mean
mean ~ dgamma(0.0001, 0.0001) #Vague hyperprior for alpha
COV ~ dgamma(0.0001, 0.0001) #Vague hyperprior for beta
inits
list(mean = 0.01, CV = 1.5)
list(mean = 0.1, CV = 0.5)
}
```

should be stable. This is clearly not the case, so we have not converged after 100,000 samples.

Running the chains for an additional 100,000 iterations gives results similar to those above. Even after more than 10^6 iterations, this behavior was still observed, and the parameter estimates provided by the two chains differed significantly.

Highly correlated parameters can be an obstacle to convergence, as noted by [8], among others. In this case, the correlation between α and β in the joint posterior distribution is estimated by OpenBUGS to be 0.98. We can ameliorate this problem by reparameterizing the gamma first-stage prior in terms of the mean and coefficient of variation, which we would expect to be less highly correlated. For a gamma distribution, the mean is given by α/β. The coefficient of variation is the ratio of the standard deviation to the mean, which for the gamma distribution will be $\alpha^{-0.5}$. We place independent, diffuse gamma hyperpriors on the mean and coefficient of variation. Initial values are again chosen using the results from the empirical Bayes analysis above. The OpenBUGS script for the reparameterized problem is shown in Table 7.10.

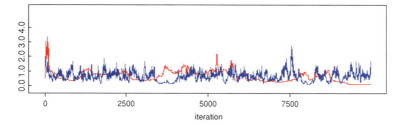

Fig. 7.14 History of 10,000 iterations for coefficient of variation, illustrating convergence

Fig. 7.15 BGR diagnostic for 10,000 iterations for coefficient of variation, illustrating convergence

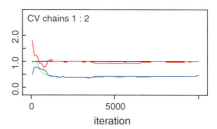

As Figs. 7.14 and 7.15 illustrate, we now appear to have convergence within the first couple of thousand iterations. To be safe, we discard the first 10,000 iterations and do not use these samples in estimating parameter values.

An additional 100,000 iterations were used to estimate the desired parameters. The predicted distribution of λ has mean 0.09/year, with a 90% credibility interval of (0.02, 0.20). These results match those from the empirical Bayes analysis above; in general, the results are not necessarily so similar. In particular, credible intervals from the hierarchical Bayesian analysis tend to be wider, especially for the overall population variability distribution (i.e., the posterior predictive distribution of λ because of the inclusion of uncertainty about the hyperparameters α and β. This uncertainty is sometimes included in empirical Bayes analyses through asymptotic approximations, although to simplify the exposition we did not do so in this example. For this example, the uncertainty in the hyperparameters is relatively small, a result of the low degree of variability among the sources.

7.4 Choice of First-Stage Prior

We now examine the impact of choosing a different functional form for the first-stage prior describing the variability in λ from source to source. We illustrate this using the data in Table 7.9. As discussed by [6], use of conjugate first-stage priors can lead to unreasonably small lower percentiles, especially in cases where there are several orders of magnitude variability among the sources being modeled.

Fig. 7.16 Marginal posterior
distribution for σ, with
"spike" at 0, an indication
of little source-to-source
variability

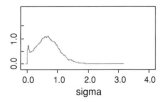

With modern tools such as OpenBUGS, conjugate distributions are no longer
a requirement of the analysis. A common nonconjugate alternative to the gamma
first-stage prior is the lognormal distribution:

$$\pi_1(\lambda | \mu, \sigma) = \frac{1}{\sqrt{2\pi}\sigma\lambda} \exp\left[-\frac{(\log\lambda - \mu)^2}{2\sigma^2}\right], \quad -\infty < \mu < \infty, \sigma > 0$$

Care must sometimes be taken in the choice of hyperpriors in order to avoid
numerical difficulties. In particular, the diffuse gamma distribution used as
a second-stage prior above can be problematic as a hyperprior for σ (and for
dispersion parameters generally). This issue is discussed at length in [1], where
a uniform distribution with finite bounds is recommended as an alternative.
A value of 1.4 for σ corresponds to a lognormal error factor of 10, and thus to
a ratio of 100 between the 95 and 5th percentiles. Therefore, a uniform (0, 5)
hyperprior for σ should be quite diffuse in practice. We will check the marginal
posterior distribution for σ to ensure that the upper tail has not been truncated,
a sign that a wider hyperprior distribution is needed.

With a uniform (0, 5) hyperprior on σ, and a flat hyperprior on μ (μ which is
a logarithmic mean, can be negative), convergence was achieved in the first 1,000
iterations. Running another 10,000 samples to estimate parameter values, we find
the mean of the posterior predictive distribution of λ to be 0.10. The 90% credible
interval is (0.02, 0.24), slightly wider than with the gamma first-stage prior above.
The marginal posterior density of σ is shown below. The right tail is not truncated,
indicating that the uniform (0, 5) hyperprior was sufficiently diffuse. Note the
"spike" at 0 in this distribution. This is an indication that there is little source-
to-source variability in the data, as noted above (Fig. 7.16).

In [9] the authors present a hierarchical Bayesian analysis of failure rate data for
digital instrumentation and control equipment. Figure 7.17 illustrates the extremely large variability in the sources the authors examined (many intervals do not
overlap), and suggests that a highly skewed distribution will be required to
adequately model the source-to-source variability.

With a gamma first-stage prior, and diffuse hyperpriors on α and β, OpenBUGS
converges within 1,000 iterations, with no reparameterization needed. An additional 50,000 iterations gave a posterior predictive mean value of 0.09, a median of
0.011, and a 90% credible interval of (6.65 $\times$ 10^{-8}, 0.43). The posterior mean of
the gamma shape parameter, α, was 0.24. Such a small value of α leads to a sharp
vertical asymptote at zero in the first-stage prior, with an excessive amount of

Fig. 7.17 Side-by-side plot of 95% intervals for data sources in [9]. Dots are posterior means from update of Jeffreys prior

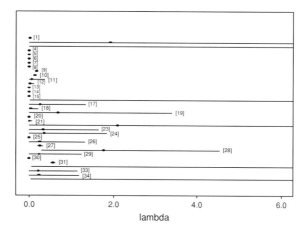

Fig. 7.18 Marginal posterior distribution for σ illustrating truncation of upper tail

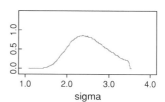

probability mass near zero, which is the only way a gamma first-stage prior can accommodate such extremely large source-to-source variability.

The lognormal distribution is perhaps a better choice for a first-stage prior to model variability that ranges over several orders of magnitude. The authors of [9] recommend a lognormal distribution in place of the gamma distribution; however, they based their hyperpriors for μ and σ on the data, using a uniform$(-7, -0.1)$ distribution for μ, and a uniform$(1, 3.5)$ distribution for σ. With these choices, the posterior predictive distribution of λ has a mean of 0.29, a median of 0.007, and a 90% interval of $(8.6 \times 10^{-5}, 0.49)$. Figure 7.18 shows the marginal posterior distribution for σ, showing the truncation of the upper tail caused by the overly narrow data-based hyperprior.

The mean of the lognormal distribution is $\exp(\mu + \sigma^2/2)$. Thus, the mean is very sensitive to σ; the median, given by e^μ, is unaffected by σ. If we replace the data-based hyperpriors used by [9] with more diffuse distributions, we expect the mean to increase. We replaced the hyperprior for μ with a flat prior over the real axis, and the distribution for σ with a more diffuse uniform$(0, 5)$ distribution. With these more diffuse hyperpriors, the mean of the predictive distribution for λ increased to 1.1, with a 90% credible interval of $(6.3 \times 10^{-5}, 0.55)$. Note that the mean, which lies well to the right of the 95th percentile, is no longer a representative value. The median, as expected, was robust against this change, remaining at 0.007. The marginal posterior distribution for σ is shown below. Note the lack of truncation in the upper tail (Fig. 7.19).

Fig. 7.19 Marginal posterior
distribution for σ with more
diffuse hyperprior illustrating
no truncation in upper tail

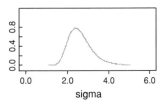

Table 7.11 OpenBUGS script for modeling plant-to-plant variability in EDG performance,
logistic-normal first-stage prior

```
model {
for (i in 1 : N) {
        x[i] ~ dbin(p[i], n[i]) #Binomial dist. for failures at each plant
        p[i] <- exp(log.p[i])/(1 + exp(log.p[i])) #First-stage prior for FTS probability at
each plant
        log.p[i] ~ dnorm(mu, tau)
        }
#Overall average distribution (posterior predictive distribution)
p.pred <- exp(log.p.pred)/(1 + exp(log.p.pred))
log.p.pred ~ dnorm(mu, tau) inits
#Hyperpriors (second-stage priors)
mu ~ dflat()
sigma ~ dunif(0, 10)
tau <- pow(sigma, -2)
}
list(mu = -5, sigma = 1)
list(mu = -7, sigma = 2)
```

To model source-to-source variability in p, the parameter of a binomial aleatory
model, one could use a lognormal distribution. However, because the range of the
lognormal distribution is unbounded, care must be exercised to ensure that values
of $p > 1$ are not produced. To avoid this difficulty, the analyst may wish to use
a logistic-normal distribution as a nonconjugate first-stage prior. The density
function of the logistic-normal distribution is given by

$$\pi_1(p|\mu, \sigma) = \frac{1}{\sqrt{2\pi}\sigma p(1-p)} exp\left\{-\frac{[ln(p/1-p) - \mu]^2}{2\sigma^2}\right\}$$

In the case of the lognormal distribution, the logarithm of the lognormally
distributed variable has a normal distribution with mean μ and variance σ^2. The
relationship is similar for the logistic-normal distribution, only it is the logit of the
variable that is normally distributed. The OpenBUGS script in Table 7.11
implements the logistic-normal hierarchical model for p, with diffuse hyperpriors
on μ and σ (recall that OpenBUGS parameterizes the underlying normal distribution in terms of $\tau = \sigma^{-2}$.

Fig. 7.20 Plot of 95%
posterior credible intervals
for λ, based on updating
Jeffreys prior with data from
Table 7.3

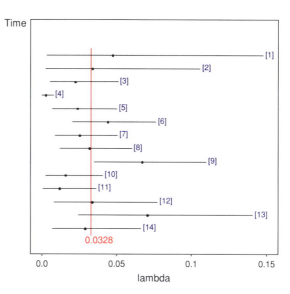

Using the data in Table 7.4, the model converges within about 1,000 iterations. Running an additional 100,000 iterations gives a posterior predicted mean for `p.pred` of 0.03, with a 90% credible interval of (6.4E-4, 012). Recall that with a beta first-stage prior, the posterior predicted mean and 90% interval were 0.02 and (3.7E-4, 0.06), so the choice of first-stage prior does have a discernible influence.

7.5 Trend Models Revisited

Let us return now to the example from [2], which we have explored both here and in Chap. 5. In Chap. 5, we saw that a simple loglinear trend model for λ did not have good predictive ability. Earlier in this chapter, we enhanced the loglinear model by introducing additional variability over time, and found that such a model had substantially better predictive ability than the simple trend model examined in Chap. 5. However, our initial graphical exploration in Chap. 5 was a caterpillar plot of λ, based on updating the Jeffreys prior for λ with the data in each time bin. This plot, reproduced in Fig. 7.20, was not suggestive of a monotonic time trend in λ, but appeared to be more indicative of random variability in λ across the time bins.

We can use the script in Table 7.8 to fit a hierarchical model representing extra-Poisson variability across the time bins, but without an underlying time trend. Doing this gives a model with good predictive ability, as shown by the posterior predictive plot in Fig. 7.21, and the Bayesian p-value of 0.51.

So we have two candidate models, both of which perform well on the posterior predictive checks: a loglinear trend model with additional variability over time and

Fig. 7.21 Plot of 95% posterior predictive intervals for replicated data in each time bin. All observed data fall within these intervals, suggesting good predictive ability of hierarchical model for extra-Poisson variability across time

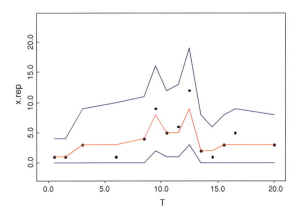

a model with no trend but with extra-Poisson variability over time. Which model do we choose?

The answer to this question depends on our assumptions about the data-generating process. Choosing the hierarchical model without a trend means that we are indifferent to re-ordering the data. In other words, we have no expectations that the event counts will be lower (or higher) in earlier time periods than in later ones. In technical terms, such an assumption is referred to as *exchangeability*. In contrast, if we are not willing to assume the data are exchangeable, perhaps because we have reasons to believe that a change in operating practice would lead to increasing event counts over time, then we should adopt a trend model.

7.6 Summary

In this chapter we have presented hierarchical Bayes models for extra variability that cannot be accounted for by simple models in which the parameter of interest is constant or is a simple monotonic function of time. The key feature of these models is that the prior distribution is specified in stages, with two stages being seen most commonly in PRA applications.

Such models have mostly been applied in PRA to account for source-to-source variability in data, and most of those applications have been approximations to full hierarchical Bayes models. The two common approximations are parametric empirical Bayes models, widely applied by the U.S. NRC, and the two-stage Bayes model of [10].

Convergence can be problematic for hierarchical Bayes models, and so care must be utilized to avoid errors caused by lack of convergence. The functional form of the first-stage prior distribution can have a significant influence on the results when modeling source-to-source variability, also. When variability among the source spans orders of magnitude, a nonconjugate lognormal or logistic-normal first-stage prior helps to avoid the extremely small lower percentiles that arise from a conjugate gamma or beta first-stage prior. However, the mean of the posterior predictive distribution may no longer be a representative value, and the analyst may wish to use the median instead.

7.7 Exercises

1. Using the data in Table 7.7, replace the gamma first-stage prior with a lognormal distribution. Use a *dflat*() hyperprior for the first parameter. For the second parameter, reparameterize in terms of σ and place a *dunif*(0, 10) hyperprior on σ.

    (a) How do the mean and median of lambda.ind compare to the results in the text? Explain.
    (b) How do the 90% intervals compare?
    (c) Any conclusions about choosing a first-stage prior?

2. Using the data in Table 7.4, with a logistic-normal first-stage prior for p, and the diffuse hyperpriors in the text, compare the resulting posterior mean and 90% credible interval for EDG #1 with the results for the beta first-stage prior given in Table 7.5.

3. The file "edg_data.txt" on the website for the text contains data for failure on demand for 195 EDGs. Recalling that the MLE of p is given by x/n, you should find that the MLE is > 0.05 for EDGs 183, 184, and 191–195. There is a desire to demonstrate that $Pr(p > 0.05) < 0.05$. In English, we want to show that we are 95% sure that EDG reliability on demand is at least 95%. If we analyze each EDG separately, using the Jeffreys prior, we will find quite a few that do not meet the criterion (i.e., too many false positives). Pooling the data would also be inappropriate, giving a very narrow credible interval for p; all of the EDGs would meet the criterion by a wide margin. We would like to get a better answer than either of these simple approaches gives by developing a hierarchical Bayes model that describes the variation in p across the 195 EDGs in the dataset.

    (a) Use OpenBUGS to analyze a hierarchical Bayes model for this data. Treat the number of failures for each EDG as binomial, with $p_i \sim$ beta(α, β). Use independent diffuse hyperpriors for α and β. We want to find which EDGs have $Pr(p > 0.05) > 0.05$.
    (b) Re-analyze this model, using a *uniform*(0, 10) hyperprior for α, and an independent *uniform*(0, 100) hyperprior for β. Does this change in priors affect your conclusions about which EDGs have $Pr(p > 0.05) > 0.05$?

4. Three plants report the following data on initiating events:

Plant	No. of events	Critical years
1	5	0.5
2	1	0.5
3	14	0.8

(a) Find a posterior mean and 90% interval for each plant, using the Jeffreys prior
     for a Poisson aleatory model, and investigate whether a model where the three
     plants have the same frequency of initiating events appears to be adequate.
(b) Perform a quantitative check of whether the frequency is the same at all three
     plants.

References

1. Gelman A (2006) Prior distributions for variance parameters in hierarchical models. Bayesian
   Anal 1(3):515–533
2. Rodionov A, Kelly D, Uwe-Klügel J (2009) Guidelines for analysis of data related to ageing
   of nuclear power plant components and systems. Joint Research Centre, Institute for Energy,
   Luxembourg: European Commission
3. Siu NO, Kelly DL (1998) Bayesian parameter estimation in probabilistic risk assessment.
   Reliab Eng Syst Saf 62:89–116
4. Kelly DL, Smith CL (2009) Bayesian inference in probabilistic risk assessment–the current
   state of the art. Reliab Eng Syst Saf 94:628–643
5. Singpurwalla ND (2006) Reliability and risk: a bayesian perspective. Wiley, Washington, DC
6. Atwood C, LaChance JL. Martz HF, Anderson DJ, Engelhardt ME, Whitehead D et al (2003)
   Handbook of parameter estimation for probabilistic risk assessment. U. S. Nuclear
   Regulatory Commission
7. Atwood CL et al (1997) Evaluation of Loss of Offsite Power Events at Nuclear Power Plants:
   1980–1996, U. S. Nucl Regulatory Comm, NUREG/CR-5496
8. Gelman A et al (2004) Bayesian data analysis, 2nd edn. Chapman and Hall/CRC, London
9. Yue M, Chu T-L (2006) Estimation of failure rates of digital components using a hierarchical
   bayesian method. International conference on probabilistic safety assessment and
   management. New Orleans
10. Kaplan S (1983) On a two-stage bayesian procedure for determining failure rates. IEEE Trans
    Power Apparatus Syst 102(1):195–262

Chapter 8
More Complex Models for Random Durations

When time is the random variable of interest, the simplest aleatory model is the exponential distribution, which was discussed in Chap. 3. For example, the exponential distribution was used to model fire suppression time in the guidance for fire PRA published jointly by the Nuclear Regulatory Commission and the Electric Power Research Institute (EPRI–NRC, 2005). However, there are numerous applications in which the exponential model, with its assumption of time-independent rate (of suppression, recovery, etc.) is not realistic. For example, past work described in [1] and [2] has shown that the rate of recovering offsite ac power at a commercial nuclear plant is often a decreasing function of time after power is lost. Therefore, the exponential distribution is not usually an appropriate model in this situation, and the analyst is led to more complex aleatory models that allow for time-dependent recovery rates, such as the Weibull and lognormal distributions.

Bayesian inference is more complicated when the likelihood function is other than exponential, and much past work has been done on various approximate approaches for these cases. In past PRAs, the difficulty of the Bayesian approach has led analysts to use frequentist methods instead, such as MLEs for point estimates and confidence intervals to represent uncertainties in the estimates. This was the approach adopted, for example, in both [1] and [2], and is the approach still in use for most commercial utility PRAs. Today, however, OpenBUGS can implement a fully Bayesian approach to the problem, allowing full propagation of parameter uncertainty through quantities derived from the aleatory model parameters, such as nonrecovery probabilities. Failure to consider parameter uncertainty can lead to nonconservatively low estimates of such quantities, and thus to overall risk metrics that are nonconservative. We will use an example of recovery of offsite ac power to illustrate the techniques in this chapter.

D. Kelly and C. Smith, *Bayesian Inference for Probabilistic Risk Assessment*,
Springer Series in Reliability Engineering, DOI: 10.1007/978-1-84996-187-5_8,
© Springer-Verlag London Limited 2011

Fig. 8.1 DAG for
exponential aleatory model

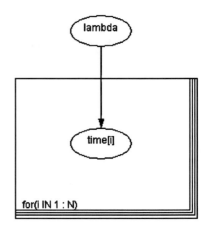

Fig. 8.1 DAG for
exponential aleatory model

8.1 Illustrative Example

Consider the following times to recover offsite power following a failure of the
offsite grid (a grid-related loss of offsite power (LOSP)). The units are hours: 0.1,
0.133, 0.183, 0.25, 0.3, 0.333, 0.333, 0.55, 0.667, 0.917, 1.5, 1.517, 2.083, 6.467.
We will analyze these data with various alternative aleatory models that allow for
a nonconstant recovery rate, and compare these models with the simpler exponential model, which has a constant recovery rate. We will use the posterior
predictive checks and information criteria described in Chap. 4 and in this Chapter
to select among the candidate aleatory models. We will also illustrate the impact of
parameter uncertainty on a derived quantity of interest, namely the probability of
not recovering the offsite grid by a critical point in time.

8.2 Analysis with Exponential Model

Recall from Chap. 3 that the simplest aleatory model for random durations is the
exponential model. Recall also the assumption that the observed times constitute a
random sample of size N from a common aleatory distribution. The exponential
distribution has one parameter, λ, which in this case is the recovery rate of the
offsite grid. This leads to the DAG shown in Fig. 8.1.

8.2.1 Frequentist Analysis

As stated above, many current PRAs have carried out a frequentist analysis for
offsite power recovery. Therefore, for comparison purposes, and because the
frequentist point estimates can be used as initial values in the Bayesian analysis,

we present some salient frequentist results. For the exponential model, the MLE is given by

$$\hat{\lambda} = \frac{N}{\sum t_i}. \tag{8.1}$$

Confidence limits for λ are given by the following equations. In these equations, $\chi^2_\alpha(d)$ is the $100 \times \alpha$th percentile of a chi-square distribution with d degrees of freedom.

$$\lambda_{\text{lower}} = \frac{\chi^2_{\alpha/2}(2N)}{2 \sum t_i}$$
$$\lambda_{\text{upper}} = \frac{\chi^2_{1-\frac{\alpha}{2}}(2N)}{2 \sum t_i} \tag{8.2}$$

For our example, the MLE of λ is 0.913/h, and the 90% confidence interval for λ is (0.55, 1.35).

8.2.2 Bayesian Analysis

To allow the observed data to drive the results, we will use the Jeffreys prior for λ. Recall from Chap. 3 that this is an improper prior, proportional to $1/\lambda$, which we can think of for purposes of Bayesian updating as a gamma distribution with both parameters equal to zero. Because the gamma prior is conjugate to the exponential likelihood function, the posterior distribution is also gamma, with parameters N and Σt_i. The posterior mean is thus $N/\Sigma t_i$. Credible intervals must be found numerically, which can be done with a spreadsheet or any basic statistics package. Because the chi-square distribution is a gamma distribution with shape parameter equal to half the degrees of freedom, and scale parameter equal to $\frac{1}{2}$, a 90% credible interval will be numerically identical to the 90% confidence interval calculated with Eq. 8.2.

Although it is not necessary to use OpenBUGS with this example, because of the conjugate nature of the prior and likelihood, we can do so. The script in Table 8.1 is the same as was used in Chap. 3, and implements the DAG shown in Fig. 8.1. Note the use of the barely proper gamma (0.0001, 0.0001) distribution to approximate the improper Jeffreys prior for λ.[1]

Note the specification of an initial value for λ, necessary because of the difficulty in initially sampling from the nearly improper prior distribution. Because this is a conjugate single-parameter problem, one Monte Carlo chain is sufficient.

---

[1] One can use a gamma (0, 0) prior in OpenBUGS, as long as initial values are loaded. However, we prefer to avoid the use of an improper prior generally, as it can lead to numerical difficulties on occasion, especially when more than one parameter is involved.

Table 8.1 OpenBUGS script for exponential aleatory model with Jeffreys prior for λ

```
model {
for(i in 1:N) {
time[i] ~ dexp(lambda) #Exponential aleatory model for recovery time
}
lambda ~ dgamma(0.0001, 0.0001) #Jeffreys prior for lambda
prob.nonrec <- exp(-lambda*time.crit)
}

data
list(time = c(0.1,0.133,0.183,0.25,0.3,0.333,0.333,0.55,
    0.667,0.917,1.5,1.517,2.083,6.467), N = 14, time.crit = 8)
inits
list(lambda = 1)
```

Using the conservative analysis criteria from Chap. 3 of 1,000 burn-in iterations, followed by 100,000 samples, we find a posterior mean for λ of 0.913/h, and a 90% credible interval of (0.55, 1.35). As expected, because we are using the Jeffreys prior and the observed random variable (time) is continuous, these are the same values, within Monte Carlo sampling error, as the frequentist estimates given above.

8.3 Analysis with Weibull Model

The Weibull distribution is an alternative aleatory model for random durations, but it has two parameters instead of one, and allows for a time-dependent recovery rate. It has a shape parameter, which we will denote as β. The second parameter, denoted α, is a scale parameter, and determines the time units. If $\beta = 1$, the Weibull distribution reduces to the exponential distribution. If $\beta > (<) 1$, the recovery rate is increasing (decreasing) with time. In the conventional parameterization, the Weibull density function is

$$f(t) = \frac{\beta}{\alpha} \left(\frac{t}{\alpha}\right)^{\beta-1} \exp\left[-\left(\frac{t}{\alpha}\right)^{\beta}\right] \tag{8.3}$$

OpenBUGS uses a slightly different parameterization, one that has been used in the past for Bayesian inference in the Weibull distribution. In this parameterization, the scale parameter is $\lambda = \alpha^{-\beta}$. In this parameterization, the density and cumulative distribution function are given by the following equations.

$$f(t) = \beta \lambda t^{\beta-1} \exp\left(-\lambda t^{\beta}\right)$$
$$F(t) = 1 - \exp\left(-\lambda t^{\beta}\right) \tag{8.4}$$

The frequentist analysis for the Weibull parameters is more complicated than for the simple exponential model. The MLEs cannot be written down in closed

Fig. 8.2 DAG for Weibull aleatory model

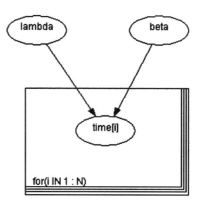

form and must be found numerically, while confidence intervals are even trickier; see [3], for example. The MLEs for β and λ for our example data are 0.84 and 1.01, respectively. These will be used to aid in selecting initial values for the Markov chains in the Bayesian analysis below.

The DAG for this model is shown in Fig. 8.2. Note that in this DAG, before the times are observed, β and λ are independent, and as a result we will use independent diffuse prior distributions. Once the data are observed, β and λ become dependent, and this is reflected in the joint posterior distribution.

The OpenBUGS script used to analyze this model is shown in Table 8.2. Note that two Monte Carlo chains are used, as there are two parameters to estimate. Convergence is very quick for this model; however, a check for convergence should always be made in models with two or more parameters. We discarded the first 1,000 iterations to be conservative, and used another 100,000 iterations to estimate β and λ. The posterior means for β and λ are 0.83 and 1.01, respectively, very close to the MLEs, as expected because of the diffuse priors. The 90% credible intervals are (0.58, 1.11) for β and (0.59, 1.525) for λ. The correlation coefficient between β and λ in the joint posterior distribution is -0.32, reflecting the dependence induced by the observed data (Table 8.2).

8.4 Analysis with Lognormal Model

A lognormal aleatory model was used for recovery of offsite power in [2]. Like the Weibull model, there are two unknown parameters, commonly denoted by μ and σ. However, unlike the Weibull distribution, neither parameter determines the shape of the lognormal distribution, as the shape is fixed. The recovery rate increases initially, and then decreases monotonically. This makes the lognormal model attractive for situations where there is a mixture of easy-to-recover and hard-to-recover events. The median is determined by μ, with the median being e^{μ}. The other parameter, σ, determines the spread of the distribution. A commonly used measure of the spread is the error factor, defined as the ratio of the 95th percentile

Table 8.2 OpenBUGS script for Weibull aleatory model with independent, diffuse priors

```
model {
for(i in 1:N) {
time[i] ~ dweib(beta, lambda) #Weibull aleatory model for recovery time
}
#Independent, diffuse priors for Weibull parameters
lambda ~ dgamma(0.0001, 0.0001)
beta ~ dgamma(0.0001, 0.0001)
}

data
list(time = c(0.1,0.133,0.183,0.25,0.3,0.333,0.333,0.55,0.667,0.917,1.5,1.517,2.083,6.467),
    N = 14)
inits
Chain 1
list(beta = 0.5, lambda = 1)
Chain 2
list(beta = 1, lambda = 0.1)
```

Fig. 8.3 DAG for lognormal aleatory model

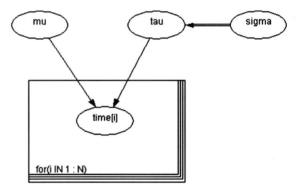

to the median. The error factor is given by $e^{1.645\sigma}$. The MLEs of μ and σ are available in closed form:

$$\hat{\mu} = \frac{1}{N} \sum \log(t_i)$$
$$\hat{\sigma} = \frac{1}{N} \sum [\log(t_i) - \hat{\mu}]^2$$

(8.5)

For the example data, the MLEs of μ and σ are −0.605 and 1.13, respectively.

OpenBUGS parameterizes the lognormal (and normal) distribution in terms of μ and the precision, τ, with $\tau = \sigma^{-2}$. The DAG for this model is shown in Fig. 8.3, and the accompanying OpenBUGS script is in Table 8.3. We use independent, diffuse priors for μ and σ, and let OpenBUGS infer the induced prior on τ. This is shown in the DAG by the double arrow connecting τ and σ, indicating the logical connection. Note the use of the *dflat()* prior for μ in the OpenBUGS script. This is an improper flat prior over the entire real axis, and is used because μ is not

Table 8.3 OpenBUGS script for lognormal aleatory model with independent, diffuse priors

```
model {
for(i in 1:N) {
time[i] ~ dlnorm(mu, tau) #Lognormal aleatory model for recovery time
}
#Independent, diffuse priors for lognormal parameters
mu ~ dflat()
sigma ~ dgamma(0.0001, 0.0001)
#Calculate tau
tau <- pow(sigma, -2)
}

 data
list(time = c(0.1,0.133,0.183,0.25,0.3,0.333,0.333,0.55,0.667,0.917,1.5,1.517,2.083,6.467),
     N = 14)
inits
Chain 1
list(mu = -0.5, sigma = 1)
Chain 2
list(mu = 0, sigma = 1.5)
```

restricted to positive values, as were the parameters in the Weibull and exponential models. We used the MLEs to determine initial values, and used two Monte Carlo chains to facilitate checking for convergence.

As with the Weibull model, convergence occurs very quickly. Conservatively discarding the first 1,000 iterations, and using another 100,000 iterations gives posterior means for μ and σ of $-$ 0.605 and 1.24, respectively, very close to the MLEs, as expected with diffuse priors. The 90% credible intervals are $(-1.16, -0.05)$ for μ and $(0.89, 1.73)$ for σ.

8.5 Analysis with Gamma Model

The gamma aleatory model is similar to the Weibull model in that it has a shape parameter, denoted by α, and a scale (really a rate) parameter, denoted by β. The density function is given by

$$f(t) = \frac{\beta^{\alpha} t^{\alpha-1} e^{-\beta t}}{\Gamma(\alpha)}$$

If $\alpha = 1$, the gamma distribution reduces to the exponential distribution. If $\alpha > (<)$ 1, the recovery rate is increasing (decreasing) with time. As with the Weibull distribution, frequentist analysis is more complicated than for the simple exponential model. The MLEs cannot be written down in closed form and must be found numerically. The MLEs for α and β for our example data are 0.85 and 0.77,

Fig. 8.4 DAG for gamma
aleatory model

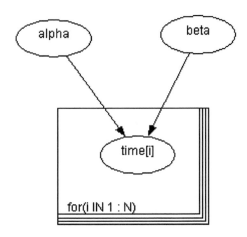

respectively. These will be used to aid in selecting initial values for the Markov chains in the Bayesian analysis below. The DAG for the gamma model is shown in Fig. 8.4. As before, we use independent diffuse priors for α and β. The OpenBUGS script is shown in Table 8.4.

Running this script as before we find the posterior mean of α and β to be 0.81 and 0.74, respectively, close to the MLEs, as expected with diffuse priors. The 90% intervals are (0.43, 1.30) and (0.28, 1.345).

8.6 Estimating Nonrecovery Probability

In a typical PRA, if offsite power is lost, the probability of not recovering power by a critical point in time is a quantity of prime interest. In general, the nonrecovery probability is given in terms of the aleatory cumulative distribution function by $1 - F(t_{crit})$. For the exponential model, this becomes

$$P_{\text{nonrec}}(t) = \exp(-\lambda t_{\text{crit}}). \tag{8.6}$$

Obviously, this probability is a function of the recovery rate, λ, and because there is uncertainty in λ, this uncertainty should be propagated through this equation to give the uncertainty in the nonrecovery probability. For the Weibull model, the nonrecovery probability is a function of both β and λ, as shown by the following equation.

$$P_{\text{nonrec}}(t) = \exp\left(-\lambda t_{\text{crit}}^{\beta}\right) \tag{8.7}$$

For the lognormal model, the nonrecovery probability is given in terms of the cumulative standard normal distribution:

Table 8.4 OpenBUGS script for gamma aleatory model with independent, diffuse priors

```
model {
for(i in 1:N) {
time[i] ~ dgamma(alpha, beta #Gamma aleatory model for recovery time
  }
#Independent, diffuse priors for gamma parameters
alpha ~ dgamma(0.0001, 0.0001)
beta ~ dgamma(0.0001, 0.0001)
}

data
list(time = c(0.1,0.133,0.183,0.25,0.3,0.333,0.333,0.55,0.667,0.917,1.5,1.517,2.083,6.467),
    N = 14)
inits
Chain 1
list(alpha = -0.5, beta = 1)
Chain 2
list(alpha = 1, beta = 0.5)
```

$$P_{\text{nonrec}}(t) = 1 - \Phi\left(\frac{\log t_{\text{crit}} - u}{\sigma}\right) \tag{8.8}$$

For the gamma model, the cumulative distribution function cannot be written in closed form. However, even in this case, the nonrecovery probability can be estimated, with full and proper treatment of parameter uncertainty, simply by adding one additional line to the OpenBUGS script for the aleatory model under consideration. Letting `time.crit` denote the critical time of concern, the following line is added to the respective scripts:

- `prob.nonrec <- 1 - cumulative(time.rep, time.crit)`

This line instructs OpenBUGS to draw samples from the posterior distribution for the relevant parameters, and for each iteration, to calculate the nonrecovery probability, using the `cumulative()` function in OpenBUGS. This constitutes Monte Carlo evaluation of the integral of the nonrecovery probability over the posterior distribution. For example, in the case of the lognormal aleatory model, this extra line of script evaluates the following integral:

$$\int_0^\infty \int_{-\infty}^\infty \left[1 - \Phi\left(\frac{\log t_{crit} - \mu}{\sigma}\right)\right] \pi_1(\mu, \sigma | t_1, t_2, \ldots, t_n) d\mu d\sigma \tag{8.9}$$

Table 8.5 compares the point estimate nonrecovery probabilities for each of the four aleatory models we have considered with the posterior means and 90% credible intervals from the full Bayesian analysis. A critical time of eight hours was used in the calculations. Note the significance of including parameter uncertainty, as well as the influence of the aleatory model.

Table 8.5 Comparison of nonrecovery probabilities at 8 h for each aleatory model

Model	Point estimate	Posterior mean	90% interval
Exponential	6.7E-4	2.8E-3	(2.0E-5, 8.1E-4)
Weibull	2.8E-3	0.012	(5.6E-5, 0.05)
Lognormal	8.6E-3	0.022	(8.9E-4, 0.08)
Gamma	1.4E-3	8.3E-3	(3.0E-5, 0.04)

8.6.1 Propagating Uncertainty in Convolution Calculations

Consider a typical sequence cut set from a LOSP event tree:

IE-LOOP * EPS-DGN-FTS-A * EPS-DGN-FTR-B * OSP-NONREC

This cut set represents the joint occurrence of the LOSP (the initiating event, IE-LOOP), the failure of emergency diesel generator (EDG) A to start (EPS-DGN-FTS-A), the failure of EDG B to run for the required mission time (EPS-DGN-FTR-B), and the failure to recover offsite power in time to prevent core damage (OSP-NONREC). Let us focus just on the time-dependent portion of the cut set, namely EPS-DGN-FTR-B * OSP-NONREC.

To quantify basic event EPS-DGN-FTR-B, the analyst must specify a mission time for the EDG; however, this is difficult, given that the EDG only needs to run until offsite power is recovered, and the time of offsite power recovery is random. Many past PRAs have dealt with this problem by specifying a surrogate EDG mission time. However, it is becoming increasingly common to treat this problem via a convolution integral:

$$P(\text{EPS-DGN-FTR-B*OSP-NONREC}) = \int_0^{t_m} \lambda e^{-\lambda t}[1 - F(t + t_c)]dt \qquad (8.10)$$

In this equation, λ is the failure rate of EDG B, and F(t) is the cumulative distribution function of the offsite power recovery time. The mission time, t_m, is the overall mission time for the sequence, which is typically 24 h. The other variable in this equation is t_c, which is the difference between the time to core damage and the time needed to restore power to the safety buses, after the grid has been recovered.

There are a number of uncertainties in this equation, both explicit and implicit. Explicitly, there is epistemic uncertainty in the EDG failure rate, λ. There is also some uncertainty in the time to core damage and the time it takes operators to restore power to the safety buses, causing t_c to be uncertain. The aleatory model for offsite power recovery time has one or more parameters, and the epistemic uncertainty in these parameters leads to uncertainty in F. Finally, there is uncertainty associated with the choice of aleatory model for offsite power recovery. In the typical PRA treatment of this equation, none of these uncertainties are addressed; only point estimate values are used.

We will use the offsite power recovery times from our earlier example to illustrate the impacts of these uncertainties. We will assume that the uncertainty in the EDG failure rate, λ, is lognormally distributed with a mean of 10^{-3}/h and an error factor of 5. For the time to core damage, denoted t_{CD}, we will assume a best estimate time of 1.5 h, with uncertainty described by a uniform distribution between 0.75 and 2 h. For the time needed by operators to restore power to the safety buses, denoted t_0, we assume a best estimate of 0.1 h, with uncertainty described by a uniform distribution between 0.05 and 0.3 h.

To illustrate the impact that uncertainty has on this problem, we first illustrate the results using point estimates for all of the parameters. We use the posterior mean values for the recovery model parameters, the mean of the EDG failure rate, and the best estimate times, with the typical PRA mission time of 24 h. Evaluating the convolution integral in Eq. 8.10 gives a core damage probability of 2.4×10^{-4} for the exponential recovery model, and 3.7×10^{-4} for the Weibull model.

To address uncertainty, we use the OpenBUGS script shown in Table 8.6. Note the use of the integral functional in OpenBUGS to carry out the convolution. Note that the variable of integration must be denoted by "x," and x is indicated as missing (NA) in the data statement. The results are shown for each model, with and without uncertainty in t_{CD} and t_0 (Tables 8.7, 8.8).

Because of the manner in which most PRA software quantifies minimal cut sets, it is common to estimate a cut set "recovery factor," which is multiplied by the remaining events in the cut set to obtain the cut set frequency. For this particular example, the recovery factor would be given by the following equation.

$$RF = \frac{P(\text{EPS-DGN-FTR-B}^*\text{OSP-NONREC})}{P(\text{EPS-DGN-FTR-B})} \tag{8.11}$$

The numerator is the core damage probability from the convolution integral, and the denominator is the probability of EDG failure to run over the PRA mission time. Using the point estimate values for the convolution integral of 2.4×10^{-4} for the exponential recovery model, and 3.7×10^{-4} for the Weibull model, with a point estimate EDG failure probability of 0.024 over a 24 h mission time, we obtain point estimate recovery factors of 0.01 and 0.016 for the exponential and Weibull models, respectively.

To consider the uncertainty in the recovery factor, we included lines in the script in Table 8.6 to carry out the calculation. Tables 8.9 and 8.10 show the results. Note the considerable uncertainty expressed by the 90% credible intervals. This is the result of the recovery factor being a ratio of two uncertain quantities. The differences from the point estimates tend to increase as the number of recovery times decreases, and failure to address uncertainty in the convolution can lead to considerable nonconservatism with sparse recovery data (see Exercise 8.5).

Table 8.6 OpenBUGS script to propagate uncertainty through convolution calculation of core damage probability

```
model {
for(i in 1:N) {
# time[i] ~ dweib(beta, lambda) #Weibull aleatory model for recovery time
time[i] ~ dexp(lambda) #Exponential aleatory model for recovery time
}
#Diffuse prior distributions
#beta ~ dgamma(0.0001, 0.0001)
lambda ~ dgamma(0.0001, 0.0001)
#Specify integrand of convolution integral (uses OpenBUGS integral functional)
#Weibull aleatory model
#F(x) <- lambda.edg*exp(-lambda.edg*x)*exp(-lambda*pow(x + T.c, beta))
#Exponential aleatory model
F(x) <- lambda.edg*exp(-lambda.edg*x)*exp(-lambda*(x + T.c))
#Time available to restore offsite power to avert core damage
#T.cd is time to core damage, T.0 is time to restore power to safety loads after recovering offsite
    power
T.c <- T.cd - T.0
#No uncertainty in T.cd and T.0
#T.cd <- 1.5
#T.0 <- 0.1
#Uncertainty in T.cd and T.0
T.cd ~ dunif(0.75, 2)
T.0 ~ dunif(0.05, 0.3)
#Lognormal distribution for EDG failure rate, mean = 1.0E-3/hr, EF = 5
lambda.edg ~ dlnorm(mu, tau)
tau <- pow(sigma, -2)
sigma <- log(5)/1.645
mu <- log(1.0E-3) - pow(sigma, 2)/2
prob.nonrec <- integral(F(x),0,24,1.E-6)
rec.factor <- prob.nonrec/prob.edg
prob.edg <- 1 - exp(-lambda.edg*24)
}
data
list(time = c(0.1,0.133,0.183,0.25,0.3,0.333,0.333,0.55,0.667,0.917,1.5,1.517,2.083,6.467),
    N = 14, x = NA)
inits
list(beta = 0.5, lambda = 1)
list(beta = 1, lambda = 0.1)
list(lambda = 1)
```

8.7 Model Checking and Selection

Chapter 4 introduced the concept of using replicate draws from the posterior predictive distribution in order to check if the model (prior + likelihood) could replicate the observed data with a reasonable probability. In the current context,

Table 8.7 Results for convolution calculation of core damage probability considering uncertainty in recovery model parameters and EDG failure rate

Recovery model	Mean probability of core damage	90% credible interval
Exponential	3.6E-4	(2.3E-5, 1.3E-3)
Weibull	5.0E-4	(3.0E-5, 1.8E-3)

Table 8.8 Results for convolution calculation of core damage probability considering uncertainty in recovery model parameters, EDG failure rate, time to core damage, and power restoration time

Recovery model	Mean probability of core damage	90% credible interval
Exponential	4.4E-4	(2.6E-5, 1.6E-3)
Weibull	5.7E-4	(3.3E-5, 2.1E-3)

Table 8.9 Results for convolution calculation of cut set recovery factor considering uncertainty in recovery model parameters and EDG failure rate

Recovery model	Mean recovery factor	90% credible interval
Exponential	0.015	(3.4E-3, 3.5E-2
Weibull	0.021	(4.3E-3, 5.4E-2)

Table 8.10 Results for convolution calculation of cut set recovery factor considering uncertainty in recovery model parameters, EDG failure rate, time to core damage, and power restoration time

Recovery model	Mean recovery factor	90% credible interval
Exponential	0.018	(3.7E-3, 4.3E-2
Weibull	0.024	(4.7E-3, 5.9E-2)

we would generate a replicate time, and compare its distribution to the observed times. If one or more of the observed times falls in a tail of this predictive distribution, there may be a problem with the model.

The replicated time is generated by adding the following lines to the respective scripts:

- Exponential: `time.rep ~ dexp(lambda)`
- Weibull: `time.rep ~ dweib(beta, lambda)`
- Lognormal: `time.rep ~ dlnorm(mu, tau)`
- Gamma: `time.rep ~ dgamma(alpha, beta)`.

OpenBUGS generates parameter values from the posterior distribution, and then generates a time from the aleatory model with the sampled parameter values. In terms of an equation, the replicate times are samples from the posterior predictive distribution, which for a lognormal aleatory model, becomes

$$f(t) = \int_0^\infty \int_{-\infty}^\infty \frac{1}{\sqrt{2\pi}\sigma t} \exp\left[-\frac{(\log t - u)^2}{2\sigma^2}\right] \pi_1(u, \sigma | t_1, t_2, \ldots, t_n) du d\sigma \quad (8.12)$$

Fig. 8.5 DAG for Weibull
aleatory model showing
replicated time from posterior
predictive distribution

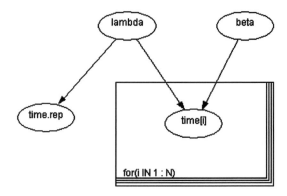

Table 8.11 Summary of
replicated times from
posterior predictive
distributions

Model	95% Credible interval
Exponential	(0.03, 4.61)
Weibull	(8.2E-3, 5.98)
Lognormal	(0.04, 7.5)
Gamma	(5.9E-3, 5.47)

In terms of a DAG, an additional node is added, as shown in Fig. 8.5 for the Weibull model.

Table 8.11 shows the 95% credible interval for the replicate time under each aleatory model. The exponential model cannot replicate the longest observed time with reasonable probability. The Weibull and gamma models do better, and the lognormal model, having the heaviest tail of the four, does the best at covering the range of observed data. However, it somewhat over-predicts the longest time.

One can also use summary statistics derived from the posterior predictive distribution. Using the Cramér-von Mises statistic, as described in [4], one can calculate a Bayesian p-value for each model, with a value near 0.5 being desirable. This statistic uses the cumulative distribution function of the ranked times, both replicated and observed. The following OpenBUGS scripts illustrate this calculation for each of the four aleatory models under consideration, using the `cumulative()` function in OpenBUGS to calculate the cumulative distribution function. The following Bayesian p-values were calculated for each of the three models: exponential, 0.42; Weibull, 0.44; lognormal, 0.58; gamma, 0.41. By this criterion, all four models are reasonable choices, with the exponential, gamma, and Weibull models tending to slightly under-predict recovery times, and the lognormal model tending to slightly over-predict (Tables 8.12, 8.13, 8.14).

It is also possible to use various information criteria, which penalize models with more parameters to protect against over-fitting. Note that these criteria are measures of relative fit, and that the best model from such a relative point of view may not be very good from an absolute point of view, so in practice it can be useful to examine replicated times and Bayesian p-value to assess model adequacy on an absolute basis. Two information criteria are discussed the Bayesian

Table 8.12 OpenBUGS script to calculate Bayesian p-value for exponential model, using Cramér-von Mises statistic from posterior predictive distribution

```
model {
for(i in 1:N) {
time[i] ~ dexp(lambda) #Exponential aleatory model for recovery time
time.rep[i] ~ dexp(lambda) #Replicate time from posterior predictive distribution
#Rank observed times and replicate times
time.ranked[i] <- ranked(time[], i)
time.rep.ranked[i] <- ranked(time.rep[], i)
#Calculate components of Cramer-von Mises statistic for observed and replicate data
F.obs[i] <- cumulative(time[i], time.ranked[i])
F.rep[i] <- cumulative(time.rep[i], time.rep.ranked[i])
diff.obs[i] <- pow(F.obs[i] - (2*i-1)/(2*N), 2)
diff.rep[i] <- pow(F.rep[i] - (2*i-1)/(2*N), 2)
}
lambda ~ dgamma(0.0001, 0.0001) #Jeffreys prior for lambda
#Calculate distribution of Cramer-von Mises statistic for observed and replicate data
CVM.obs <- sum(diff.obs[])
CVM.rep <- sum(diff.rep[])
p.value <- step(CVM.rep - CVM.obs) #Mean value should be near 0.5
}
data
list(time = c(0.1,0.133,0.183,0.25,0.3,0.333,0.333,0.55,0.667,0.917,1.5,1.517,2.083,6.467),
    N = 14)
inits
list(lambda = 1)
```

information criterion (BIC) and the deviance information criterion (DIC). BIC has been recommended for use in selecting among non-hierarchical models, while the DIC has been recommended for selecting among hierarchical models. However, in practice, we have found that a non-hierarchical model selected on the basis of BIC also tends to be the model selected on the basis of DIC (but not vice versa). Since no additional scripting is needed to calculate DIC in OpenBUGS, it is probably the easier of the two criteria to use. However, we will illustrate how to calculate BIC, also.

For BIC, we have

$$\text{BIC} = -2 \times \log \text{likelihood} + k \log N. \tag{8.13}$$

In this equation, k is the number of unknown parameters (one for the exponential model, two for the Weibull and lognormal models) and N is the number of data points (14 in our example). The model with the smallest BIC is preferred. Note that, some references swap the $+$ and $-$ signs in Eq. 8.13; under this alternative definition one selects the model with the *largest* BIC. The scripts below illustrate the calculation of BIC in OpenBUGS for each of the four aleatory models under consideration. In our example, the estimated BICs are 34.2, 36.99, 33.6, and 37.7 for the exponential, Weibull, lognormal, and gamma models, respectively.

Table 8.13 OpenBUGS script to calculate Bayesian p-value for Weibull model, using Cramér-von Mises statistic from posterior predictive distribution

```
model {
for(i in 1:N) {
time[i] ~ dweib(beta, lambda) #Weibull aleatory model
time.rep[i] ~ dweib(beta, lambda)
#Rank observed times and replicate times
time.ranked[i] <- ranked(time[], i)
time.rep.ranked[i] <- ranked(time.rep[], i)
#Calculate components of Cramer-von Mises statistic for observed and replicate data
F.obs[i] <- cumulative(time[i], time.ranked[i])
F.rep[i] <- cumulative(time.rep[i], time.rep.ranked[i])
diff.obs[i] <- pow(F.obs[i] - (2*i-1)/(2*N), 2)
diff.rep[i] <- pow(F.rep[i] - (2*i-1)/(2*N), 2)
}
#Diffuse prior distributions
lambda ~ dgamma(0.0001, 0.0001)
beta ~ dgamma(0.0001, 0.0001)
#Calculate distribution of Cramer-von Mises statistic for observed and replicate data
CVM.obs <- sum(diff.obs[])
CVM.rep <- sum(diff.rep[])
p.value <- step(CVM.rep - CVM.obs) #Mean value should be near 0.5
}
data
list(time = c(0.1,0.133,0.183,0.25,0.3,0.333,0.333,0.55,0.667,0.917,1.5,1.517,2.083,6.467),
    N = 14)
inits
#Weibull model (two chains)
list(beta = 0.5, lambda = 1)
list(beta = 1, lambda = 0.1)
```

So based on BIC, we would select the lognormal model over the other three (Tables 8.15, 8.16, 8.17, 8.18).

We now will calculate the DIC for each of the candidate models. The DIC is calculated automatically by OpenBUGS, with no need for additional scripting. The lognormal model also has the smallest DIC, 30.16 versus 32.57, 33.73, and 34.46 for the exponential, Weibull, and gamma models, so it would be selected under this criterion, also.

In summary, among the four candidate aleatory models for recovery time, the lognormal model would be chosen based on penalized likelihood criteria such as BIC or DIC. In terms of replicated recovery times, the exponential, gamma, and Weibull models tend to under-predict recovery times, with the exponential model being the worst in this regard. The lognormal model gives the best coverage of the observed recovery times, but tends to slightly over-predict these times, as indicated by the Bayesian p-value being slightly > 0.5. The choice of aleatory model can make a considerable difference in the estimated nonrecovery probability,

Table 8.14 OpenBUGS script to calculate Bayesian p-value for lognormal model, using Cramér-von Mises statistic from posterior predictive distribution

```
model {
for(i in 1:N) {
time[i] ~ dlnorm(mu, tau) #Lognormal aleatory model
time.rep[i] ~ dlnorm(mu, tau)
#Rank observed times and replicate times
time.ranked[i] <- ranked(time[], i)
time.rep.ranked[i] <- ranked(time.rep[], i)
#Calculate components of Cramer-von Mises statistic for observed and replicate data
F.obs[i] <- cumulative(time[i], time.ranked[i])
F.rep[i] <- cumulative(time.rep[i], time.rep.ranked[i])
diff.obs[i] <- pow(F.obs[i] - (2*i-1)/(2*N), 2)
diff.rep[i] <- pow(F.rep[i] - (2*i-1)/(2*N), 2)
}
#Diffuse prior distributions
mu ~ dflat()
sigma ~ dgamma(0.0001, 0.0001)
tau <- pow(sigma, -2)
#Calculate distribution of Cramer-von Mises statistic for observed and replicate data
CVM.obs <- sum(diff.obs[])
CVM.rep <- sum(diff.rep[])
p.value <- step(CVM.rep - CVM.obs) #Mean value should be near 0.5
}
data
list(time = c(0.1,0.133,0.183,0.25,0.3,0.333,0.333,0.55,0.667,0.917,1.5,1.517,2.083,6.467),
    N = 14)
inits
#Lognormal model (two chains)
list(mu = -0.5, sigma = 1.1)
list(mu = -1, sigma = 1.4)
```

especially when parameter uncertainty is accounted for properly in a fully Bayesian analysis. From this consideration, the exponential model gives the most nonconservative result in this example, because of its tendency to under-predict recovery times. The lognormal model is the most conservative of the four from this regard, because of its tendency to slightly over-predict recovery times.

8.8 Exercises

1. Reference [5] The following projector lamp failure times (in hours) have been collected: 387, 182, 244, 600, 627, 332, 418, 300, 798, 584, 660, 39, 274, 174, 50, 34, 1895, 158, 974, 345, 1755, 1752, 473, 81, 954, 1407, 230, 464, 380, 131, 1205.

Table 8.15 OpenBUGS script to calculate BIC for the exponential aleatory model

```
model {
for(i in 1:N) {
time[i] ~ dexp(lambda) #Exponential aleatory model for recovery time
#Exponential log-likelihood components
log.like[i] <- log(lambda) - lambda*time[i]
}
log.like.tot <- sum(log.like[])
#Calculate Bayesian information criterion
BIC <- -2*log.like.tot + log(N)
lambda ~ dgamma(0.0001, 0.0001) #Jeffreys prior for lambda
}
data
list(time = c(0.1,0.133,0.183,0.25,0.3,0.333,0.333,0.55,0.667,0.917,1.5,1.517,2.083,6.467),
    N = 14)
inits
list(lambda = 1)
```

Table 8.16 OpenBUGS script to calculate BIC for the Weibull aleatory model

```
model {
for(i in 1:N) {
time[i] ~ dweib(beta, lambda) #Weibull aleatory model for recovery time
#Weibull log-likelihood components
log.like[i] <- log(lambda) + log(beta) + (beta-1)*log(time[i]) - lambda*pow(time[i], beta)
}
log.like.tot <- sum(log.like[])
#Calculate Bayesian information criterion
BIC <- -2*log.like.tot + 2*log(N)
#Independent, diffuse priors for Weibull parameters
lambda ~ dgamma(0.0001, 0.0001)
beta ~ dgamma(0.0001, 0.0001)
}
data
list(time = c(0.1,0.133,0.183,0.25,0.3,0.333,0.333,0.55,0.667,0.917,1.5,1.517,2.083,6.467),
    N = 14)
inits
Chain 1
list(beta = 0.5, lambda = 1)
Chain 2
list(beta = 1, lambda = 0.1)
```

Table 8.17 OpenBUGS script to calculate BIC for the lognormal aleatory model

```
model {
for(i in 1:N) {
time[i] ~ dlnorm(mu, tau) #Lognormal aleatory model for recovery time
#Lognormal log-likelihood components
log.like[i] <- -0.5*(log(2) + log(3.14159)) - log(sigma) - log(time[i]) - pow(log(time[i])-mu, 2)/
    (2*pow(sigma, 2))
}
log.like.tot <- sum(log.like[])
#Calculate Bayesian information criterion
BIC <- -2*log.like.tot + log(N)*2
#Independent, diffuse priors for lognormal parameters
mu ~ dflat()
sigma ~ dgamma(0.0001, 0.0001)
#Calculate tau
tau <- pow(sigma, -2)
}
data
list(time = c(0.1,0.133,0.183,0.25,0.3,0.333,0.333,0.55,0.667,0.917,1.5,1.517,2.083,6.467),
    N = 14)
inits
Chain 1
list(mu = -0.5, sigma = 1)
Chain 2
list(mu = 0, sigma = 1.5)
```

a. Use a Weibull aleatory model for the failure time, with diffuse priors on the Weibull parameters. What is the posterior probability that the Weibull shape parameter exceeds 1? What does this suggest about the viability of the Weibull model compared with the exponential model?

b. Use DIC to compare exponential, Weibull, and lognormal aleatory failure time models for the lamp.

2. The following repair times have been observed. Assuming these are a random sample from an exponential aleatory model, update the Jeffreys prior for λ to find the posterior mean and 90% credible interval for λ. 105, 1, 1263, 72, 37, 814, 1.5, 211, 330, 7929, 296, 1, 120, 1.

3. Answer the questions below for the following times: 1.8, 0.34, 0.23, 0.55, 3, 1.1, 0.68, 0.45, 0.59, 4.5, 0.56, 1.6, 1.8, 0.42, 1.6.

a. Assuming an exponential aleatory model, and a lognormal prior for λ with a mean of 1 and error factor of 3, find the posterior mean and 90% interval for λ.

b. Assume a lognormal aleatory model with independent, diffuse priors on the lognormal parameters. Find the posterior mean and 90% interval for λ.

Table 8.18 OpenBUGS script to calculate BIC for the gamma aleatory model

```
model {
for(i in 1:N) {
time[i] ~ dgamma(alpha, beta) #Gamma aleatory model for recovery time
#Gamma log-likelihood components
log.like[i] <-alpha*log(beta) + (alpha-1)*log(time[i]) - beta*time[i] - loggam(alpha)
}
log.like.tot <- sum(log.like[])
#Calculate Bayeian information criterion
BIC <- -2*log.like.tot + 2*log(N)
#Independent, diffuse priors for Weibull parameters
alpha ~ dgamma(0.0001, 0.0001)
beta ~ dgamma(0.0001, 0.0001)
}
data
list(time = c(0.1,0.133,0.183,0.25,0.3,0.333,0.333,0.55,0.667,0.917,1.5,1.517,2.083,6.467),
    N = 14)
inits
Chain 1
list(alpha = 0.5, beta = 1)
Chain 2
list(alpha = 1, beta = 0.5)
```

4. The following are emergency diesel generator (EDG) repair times, in hours. 50, 4, 2, 56, 4.7, 3.7, 33, 1.1, 3.1, 13, 2.6, 12, 11, 51, 17, 28, 20, 49.

   a. Find the posterior means of the parameters of a Weibull aleatory model for these repair times. Use independent, diffuse priors for the Weibull parameters.

   b. Generate two, independent *predicted* repair times from the posterior distribution. What is the mean and 90% credible interval for each of these times?

   c. Define a new variable that is the *minimum* of each of the two times generated in part (b). What is the mean and 90% credible interval for this minimum time?

5. From past experience, we judge that the exponential distribution is not likely to be a reasonable model for duration of LOSP. This is because the recovery rate tends to decrease with time after the event. We decide to use an alternative model that allows for a decreasing recovery rate. The following 10 recovery times are observed, in hours: 46.7, 1581, 33.7, 317.1, 288.3, 2.38, 102.9, 113, 751.6, 26.4.

    a.   Use OpenBUGS to fit a Weibull distribution to these times. Use diffuse, independent hyperpriors for the two Weibull parameters. Find the posterior mean of the shape parameter. What is the probability that the recovery rate is decreasing with time?

    b.   Now use OpenBUGS to fit a lognormal model to the same times. Again, use diffuse, independent hyperpriors for the two lognormal parameters.

    c.   Use graphical and quantitative posterior predictive checks to decide if one of these models fits significantly better than the other.

    d.   Compare DIC for these models with that of a simpler exponential model.

6. Nine failure times were observed for a heat exchanger in a gasoline refinery (in year): 0.41, 0.58, 0.75, 0.83, 1.00, 1.08, 1.17, 1.25, 1.35. The analyst proposes a Weibull(β, λ) aleatory model for these failure times. The analyst believes that $\beta = 3.5$ and manufacturer reliability specifications were translated into a uniform (0.5, 1.5) prior distribution for λ. Find the posterior mean and 90% credible interval for λ. Are there any problems with replicating the observed data with this model?

7. In Ex. 6, find a 90% posterior interval for the MTTF of the heat exchanger.

References

1. Atwood CL et al (1997) Evaluation of Loss of Offsite Power Events at Nuclear Power Plants: 1980–1996, U. S. Nucl Regulatory Comm, NUREG/CR-5496
2. Eide S et al (2005) Reevaluation of station blackout risk at nuclear power plants, U. S. Nuclear Regulatory Commission, NUREG/CR-6890
3. Bain L, Engelhardt M (1991) Statistical theory of reliability and life-testing models. Marcel-Dekker, NY
4. Kelly DL, Smith CL (2009) Bayesian inference in probabilistic risk assessment—the current state of the art. Reliab Eng Syst Saf 94:628–643
5. Hamada MS, Wilson AG, Reese CS, Martz HF (2008) Bayesian reliability. Springer, New York

Chapter 9
Modeling Failure with Repair

In Chaps. 3 and 8 we analyzed times to occurrence of an event of interest. Such an event could be failure of a component or system. If the failure is not repaired, but the component or system is replaced following failure, then the earlier analysis methods are applicable. However, in this chapter, we consider the case in which the failed component or system is repaired and placed back into service. Analysis in this situation is a bit more complicated. The details will depend principally upon the nature of the system or component after repair. We will consider two cases[1]:

- Repair leaves the component or system the same as new,
- Repair leaves the component or system the same as old.

Both of these cases are modeling assumptions that an analyst must make, and they lead to different aleatory models for the failure time. We will provide some qualitative guidance for when each assumption might be appropriate, and we will provide some qualitative and quantitative model checks that can be used to check the reasonableness of the assumption.

In all of the following discussion, we assume that we can ignore the time it takes to actually repair a component or system that has failed. This allows us to treat the failure process as a simple point process. This assumption is typically reasonable either because repair time is short with respect to operational time, or because we are only concerned with operational time, so time out for repair is accounted for through component or system maintenance unavailability estimates.

---

[1] Intermediate cases, in which repair leaves the component in a state in between new and old, can also be modeled, along with imperfect repair, which leaves a component worse than old. Such more general models are less commonly used, and are still an area of research and thus practical guidelines are difficult to give, so they are not included herein.

D. Kelly and C. Smith, *Bayesian Inference for Probabilistic Risk Assessment*,
Springer Series in Reliability Engineering, DOI: 10.1007/978-1-84996-187-5_9,
© Springer-Verlag London Limited 2011

9.1 Repair Same as New: Renewal Process

In this case, repair leaves the failed component or system in the same state as a new component or system. The times between failures are thus independent and will be assumed to be identically distributed. If the times between failures are exponentially distributed, the methods of Chap. 3 can be applied. If the times between failures are not exponentially distributed (e.g., Weibull or lognormal), then the methods of Chap. 8 can be applied. Note in both cases that it is the *times between failures* that are analyzed, *not* the cumulative failure times.

The assumption of repair same as new is plausible when the entire component or system is replaced or completely overhauled following failure. Examples would be replacement of a failed circuit board or rewinding a motor.

9.1.1 Graphical Check for Time Dependence of Failure Rate in Renewal Process

If one assumes a renewal process, then a qualitative check on whether the failure rate is constant can be done using the times between failures. If the failure rate is constant in a renewal process, then the times between failures are exponentially distributed. If one plots the ranked times between failures on the x-axis, and $1/n_t$ on the y-axis, where n_t is the number of components still operating at time t, the result should be approximately a straight line if the failure rate is constant. If the slope is increasing (decreasing) with time, this suggests a renewal process whose failure rate is likewise increasing (decreasing) with time. Such a plot is referred to as a *cumulative hazard plot*.

Consider the following 25 cumulative times of failure (in days) for a servo motor, taken from [1].

There are 25 times so the cumulative hazard plot increases by 1/25 at 27.1964, by 1/24 at 74.28, etc. The plot is shown in Fig. 9.1, and appears to indicate an increasing failure rate with time. Quantitative analysis of these times can be carried out using the methods for exponential durations in Chap. 3. A quantitative check will be described below, after we have discussed the other extreme: repair same as old.

9.2 Repair Same as Old: Nonhomogeneous Poisson Process

In the case where repair only leaves the component in the condition it was in immediately preceding failure, then the times between failures may not be independent. For example, if the component is wearing out over time (aging), then later times between failures will tend to be shorter than earlier times, and conversely

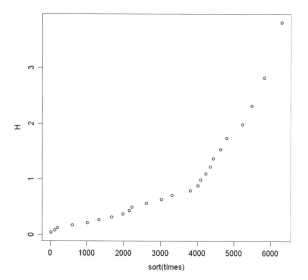

Fig. 9.1 Cumulative hazard plot for data in Table 9.1, suggesting increasing failure rate with operating time

if the component is experiencing reliability growth. In these cases, the times between failures will also fail to meet the assumption of being identically distributed. Recall that to apply the methods of Chap. 3, these two assumptions must be met.

Repair to a state that is the same as old is a reasonable default assumption for most components in a risk assessment, because a typical component is composed of subcomponents. When failure occurs, only a portion of the component (one or more subcomponents) is typically repaired, so the majority of the subcomponents are left in the condition they were in at the time of failure.

9.2.1 Graphical Check for Trend in Rate of Occurrence of Failure When Repair is same as Old

To avoid the confusion that can arise from using the same terminology in different contexts, we will follow [2] and use rate of occurrence of failure (ROCOF) instead of failure rate for the case where we have repair same as old. If the ROCOF is constant with time, then the times between failures will not tend to get shorter (aging) or longer (reliability growth) over time. If one plots the cumulative number of failures on the y-axis versus cumulative failure time on the x-axis, the resulting plot will be approximately a straight line if the ROCOF is constant. If aging is occurring (increasing ROCOF), the slope will increase with time, as the times between failures get shorter. If reliability growth is occurring (decreasing ROCOF), the slope will decrease with time, as the times between failures get longer.

Consider the following 25 cumulative times in standby at which a cooling unit failed, taken from [1]. Because the cooling unit consists of a large number of subcomponents, and only one or two of these were replaced at each failure, assume

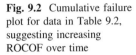

Fig. 9.2 Cumulative failure
plot for data in Table 9.2,
suggesting increasing
ROCOF over time

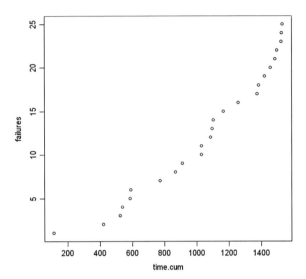

repair leaves the cooling unit in the state it was in immediately prior to failure. A plot of the cumulative number of failures versus cumulative failure time can be used to check if there appears to be a time trend in the ROCOF for the cooling unit. The plot is shown in Fig. 9.2. The slope appears to be increasing with time, suggesting that the ROCOF for the cooling unit is increasing as a function of calendar time.

An interesting exercise is to construct a cumulative failure plot for the data in Table 9.1, where the failure rate appeared to be an increasing function of operating time under the assumption that repair was same as new, a renewal process. The plot, shown in Fig. 9.3, does not suggest an increasing ROCOF under the assumption of repair same as old. This illustrates a subtle point in analyzing repairable systems. If repair is same as new after each failure, then times between failures will not exhibit a trend over *calendar* time. Aging or reliability growth only occurs over the time between one failure and the next, because the system returns to new, and the clock is reset, after each failure. Metaphorically, the system is reincarnated after each failure. On the other hand, when repair is same as old after each failure, then aging or reliability growth occurs over calendar time, and one can then expect to see a trend in the slope of cumulative failures versus time. Therefore, absence of a trend in the cumulative failure plot may suggest no aging or reliability growth under the assumption of repair same as old, but there still may be aging or reliability growth between each failure under the assumption of repair same as new. The cumulative hazard plot shown earlier can be used to check for that possibility. Metaphorically, under repair same-as-old, the system is resuscitated after each failure.

Figures 9.4 and 9.5 show cumulative failures versus cumulative time for 1,000 simulated failure times from two different renewal processes, one in which failure rate is decreasing with increasing operating time, the other where failure rate

Table 9.1 Cumulative failure times of a servo motor, from [1]	Cumulative time (days)	Time between failures (days)	Sorted time between failures
	127.4920	127.492	27.1964
	154.6884	27.1964	74.28
	330.4580	175.7696	76.22
	739.9158	409.4578	87.037
	1153.074	413.1582	123.803
	1470.720	317.646	127.492
	1809.616	338.896	141.964
	2118.147	308.531	160.479
	2289.570	171.423	171.423
	2365.790	76.22	175.7696
	2757.970	392.18	193.382
	3154.409	396.439	201.649
	3448.874	294.465	258.093
	3941.777	492.903	294.465
	4143.426	201.649	308.531
	4217.706	74.28	317.646
	4359.670	141.964	338.896
	4483.473	123.803	352.491
	4570.51	87.037	392.18
	4763.892	193.382	396.439
	4924.371	160.479	409.4578
	5360.967	436.596	413.1582
	5619.06	258.093	436.596
	5971.551	352.491	476.511
	6448.062	476.511	492.903

is increasing with operating time. Note in both cases that the cumulative failure plot produces a straight line, reinforcing the conclusion that this plot is useful for checking for aging or reliability growth under the same-as-old assumption for repair, but it cannot detect a time-dependent failure rate under the same-as-new repair assumption. The corresponding cumulative hazard plots in Figs. 9.6 and 9.7 are useful for this purpose when repair is same as new.

9.2.2 Bayesian Inference Under Same-as-Old Repair Assumption

As stated above, if there is an increasing or decreasing trend in the ROCOF over time, then the times between failures will not be independently and identically distributed, and thus the methods of Chaps. 3 and 8, which rely on this condition, cannot be applied. In particular, one should not simply fit a distribution (Weibull, gamma, etc.) to the cumulative failure times or the times between failures. Instead, the likelihood function must be constructed using the fact that each failure time, after the first, is dependent upon the preceding failure time. One must also specify a functional form for the ROCOF. We will assume a power-law form for our

Table 9.2 Twenty five cumulative times in standby at which a cooling unit failed, taken from [1]

Cumulative time (days)
116.0454
420.8451
523.1398
538.3135
585.581
591.5301
772.365
868.7294
912.3777
1031.021
1031.133
1086.673
1096.476
1103.463
1165.640
1257.554
1375.917
1385.808
1421.459
1456.259
1484.755
1496.982
1523.915
1526.050
1530.836

Fig. 9.3 Cumulative failure plot for data in Table 9.1, showing lack of trend in slope over *calendar time*

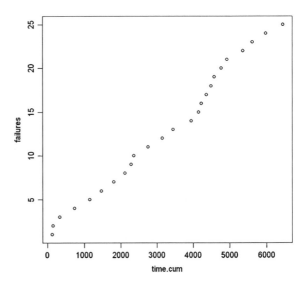

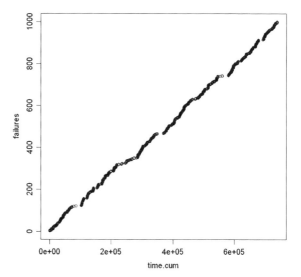

Fig. 9.4 Cumulative failure plot for 1,000 simulated failure times from renewal process with *decreasing* failure rate

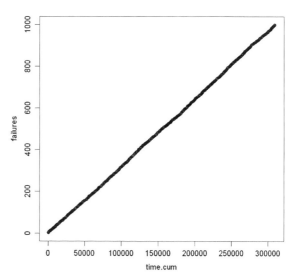

Fig. 9.5 Cumulative failure plot for 1,000 simulated failure times from renewal process with *increasing* failure rate

analysis here, as this is a commonly used aleatory model, sometimes referred to in the reliability literature as the Crow-AMSAA model. The equation for the power-law form of the ROCOF is

$$\lambda(t) = \frac{\beta}{\alpha} \left(\frac{t}{\alpha}\right)^{\beta-1} \quad \alpha, \beta > 0$$

In this model, there are two unknown parameters, which we denote as α and β. β determines how the ROCOF changes over time, and α sets the units with which time is measured. If $\beta < 1$, reliability growth is occurring, if $\beta > 1$, aging is taking place, and if $\beta = 1$, there is no trend over time.

Fig. 9.6 Cumulative hazard
plot for 1,000 simulated
failure times from renewal
process with *decreasing*
failure rate

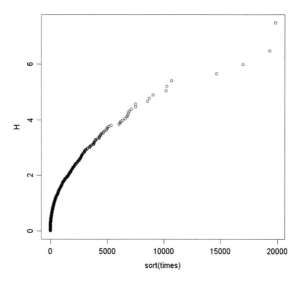

Fig. 9.7 Cumulative hazard
plot for 1,000 simulated
failure times from renewal
process with *increasing*
failure rate

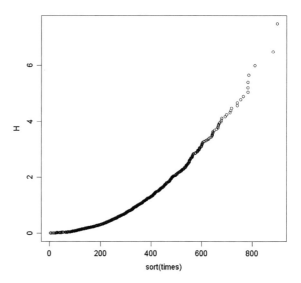

If the process is failure-truncated, and t_i is the cumulative time until the ith
failure, then the likelihood function is given by

$$f(t_1, t_2, \ldots, t_n | \alpha, \beta) = \frac{\beta^n}{\alpha^{n\beta}} \prod_{i=1}^{n} t_i^{\beta-1} \exp\left[-\left(\frac{t_n}{\alpha}\right)^\beta\right] \qquad (9.1)$$

This equation can be derived from the fact that the time to first failure has
a Weibull (β, α) distribution, and each succeeding cumulative failure time has
a Weibull distribution, truncated on the left at the preceding failure time.

Fig. 9.8 DAG for modeling
failure with repair under
same-as-old repair
assumption

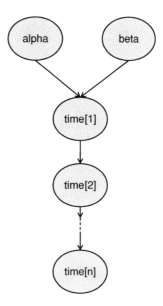

The underlying DAG model is shown in Fig. 9.8 and the OpenBUGS script is listed in Table 9.3. Model checking in this case uses the Bayesian analog of the Cramer-von Mises statistic described in Chap. 4, and is based on the fact that, for the power-law process, if one defines $z_i = (t_i/\alpha)^\beta$, then $z_{inc,i} = z_i - z_{i-1}$ has an exponential distribution with rate 1, see [3].

We run three chains. Convergence to the joint posterior distribution appears to occur within the first 1,000 samples, so we discard the first 1,000 samples for burn in. We ran another 10,000 samples to estimate parameter values, obtaining a posterior mean for β of 1.94 with a 90% credible interval of (1.345, 2.65). The posterior probability that $\beta > 1$ is near unity, suggesting a ROCOF that is increasing with time, corresponding to aging, and agreeing with the cumulative failure plot in Fig. 9.2. The Bayesian p-value is 0.57, close to the ideal value of 0.5

The script in Table 9.3 analyzes a process that is *failure-truncated*, that is, observation stops after the last failure. It is also possible to have a *time-truncated* process, in which observation continues after the last failure, up to the stopping time τ. To handle this case, only one line in the OpenBUGS script needs to be changed. The line defining the log-likelihood changes to the following:

```
for(j in 1:M) {
    phi[j] <- -log(beta) + beta*log(alpha) − (beta-1)*log(t[j]) + pow(tau/alpha, beta)/M
    }
```

The stopping time (tau) is loaded as part of the observed data.

Table 9.3 OpenBUGS script for analyzing data under same-as-old repair assumption (power-law process)

```
model {
for(i in 1:M)    {
    zeros[i] <- 0
    zeros[i] ~ dloglik(phi[i])
    #phi[i] = log(likelihood[i])
    }
#Power-law model (failure-truncated)
t[M] ~ dgamma(0.0001, 0.0001)C(t[M-1],) #Monitor node t[M] for predicted time of next
failure
for(j in 1:M)    {
    phi[j] <- log(beta) - beta*log(alpha) + (beta-1)*log(t[j]) - pow(t[M]/alpha, beta)/M
    }
#Model validation section
for(j in 1:M) {
    z.obs[j] <- pow(t[j]/alpha, beta)
    }
z.inc.obs[1] <- z.obs[1]
for(k in 2:M) {
    z.inc.obs[k] <- z.obs[k] - z.obs[k-1]
    }
for(j in 1:M) {
    z.inc.rep[j] ~ dexp(1)
    z.rep.ranked[j] <- ranked(z.inc.rep[], j)
    z.obs.ranked[j] <- ranked(z.inc.obs[], j)
    F.obs[j] <- cumulative(-z.obs.ranked[j])
    F.rep[j] <- cumulative(-z.rep.ranked[j])
    diff.obs[j] <- pow(F.obs[j] - (2*j-1)/(2*M), 2)
    diff.rep[j] <- pow(F.rep[j] - (2*j-1)/(2*M), 2)
    }
CVM.obs <- sum(diff.obs[])
CVM.rep <- sum(diff.rep[])
p.value <- step(CVM.rep - CVM.obs)
alpha ~ dgamma(0.0001, 0.0001)
beta ~ dgamma(0.0001, 0.0001)
}
Data
Load data from external file
inits
list(beta=1, alpha=100) #Inits for NHPP model
list(beta=2, alpha=50)
list(beta=0.5, alpha=200)
```

9.3 Incorporating Results into PRA

How quantitative results are incorporated into a PRA model depends again on the process assumed to describe the failures. If the repair is assumed to be same-as-new, leading to a renewal process, then the aleatory model for failure time in the PRA is the renewal distribution (e.g., Weibull). Most PRA software packages can only model an exponential distribution for failure time, so this would appear to be a problem. A way around this problem for operating equipment with a Weibull renewal distribution is as follows. Assume the mission time for the operating equipment is t_m, and assume we have determined that failures of the operating equipment can be described by a renewal process with a Weibull distribution for times between failures, with shape β and scale λ. That is, the density function for times between failures is given by

$$f(t) = \beta \lambda t^{\beta-1} \exp\left(-\lambda t^{\beta}\right)$$

This is the aleatory model for failure, which as noted above most PRA software does not include as an option. The event of interest in the PRA is the probability that the equipment fails before the end of the mission time. To have the PRA software calculate this probability, one can input an exponential distribution with rate λ as the stochastic model in the PRA, but replace the mission time by $(t_m)^{\beta}$. This is adequate for a point estimate, but may not allow epistemic uncertainty in β and λ to be propagated through the PRA model. If the renewal distribution is gamma or lognormal, there is no such simple work-around. If there is no clear preference for one model over the other (i.e., lognormal or gamma clearly better than Weibull), it is best to use a Weibull renewal distribution, as it will be easiest to incorporate into the PRA.

If the failures are assumed to be described by a power-law NHPP (repair same as old), the process for incorporating the results into the PRA is more complicated. If the component at hand is new, and its failures are assumed to be described by an NHPP with parameters estimated from past data, then the time to first failure of the component is the variable of interest, and this time will be Weibull-distributed with shape parameter β and scale parameter λ, where λ is given in terms of the power-law parameters by $\alpha^{-\beta}$. See above for how to "trick" the PRA software into using a Weibull stochastic model for time to first failure.

If the parameters have been estimated from a component's failure history, and this history continues into the future, then one is interested in the probability that the next failure will occur before the end of the mission. The distribution of the next cumulative failure time, t_i, is *not* a simple Weibull distribution: the cumulative distribution function is instead a left-truncated Weibull distribution. The probability that the next failure will occur before (cumulative) time t, given that the last failure occurred at (cumulative) time T, is given by

$$F(t) = 1 - \exp\{-\lambda[(T+t)^{\beta} - T^{\beta}]\}$$

where $\lambda = \alpha^{-\beta}$. Adding the lines below to the OpenBUGS script in Table 9.3 encodes this approach.

```
t.miss <- 24
t.window <- t[M] + t.miss
prob.fail <- 1 - exp(-pow(alpha, -beta)*(pow(t.window, beta) - pow(t[M], beta)))
```

Running the script as above, we find a mean probability of failure over a 24 h mission time of 0.53, with a 90% interval of (0.36, 0.69). One item to note is that the parameters of the power-law process, α and β, are highly correlated; the rank correlation between them is about 0.95 for the analysis above. This correlation must be taken into account in the calculation of the failure probability; failure to take it into account leads to an over-estimate of the uncertainty. With no correlation between α and β, the 90% interval for the failure probability over a 24 h mission time is estimated to be (0.16, 0.99), much wider than the interval obtained using OpenBUGS, which takes the correlation into account automatically.

9.4 Exercises

1. The following times in min are assumed to be a random sample from an exponential distribution: 1.7, 1.8, 1.9, 5.8, 10.0, 11.3, 14.3, 16.6, 19.4 and 54.8. Plot the cumulative hazard function for these times. Do the times appear to be exponentially distributed?
2. Derive Eq. 9.1.

References

1. Rodionov A, Kelly D, Uwe-Klügel J (2009) Guidelines for analysis of data related to ageing of nuclear power plant components and systems. Joint Research Centre, Institute for Energy, Luxembourg: European Commission
2. Ascher H, Feingold H (1984) Repairable systems reliability: modeling, inference, misconceptions and their causes. Marcel Dekker Inc, New York
3. Bain L, Engelhardt M (1991) Statistical theory of reliability and life-testing models. Marcel Dekker, New York

Chapter 10
Bayesian Treatment of Uncertain Data

So far we have only analyzed cases where the observed data were complete and known with certainty. Reality is messier in a number of ways with respect to observed data. For example, the number of binomial demands may not have been recorded and thus have to be estimated. Similarly, the exposure time in the Poisson distribution may have to be estimated. One may not always be able to ascertain the exact number of failures that have occurred, perhaps because of imprecision in the failure criterion. When observing times at which failures occur (i.e., random durations), various types of *censoring* can occur, leading to incomplete data. For example, a number of components may be placed in test, but the test is terminated before all the components have failed. This produces a set of observed data consisting of the recorded failure times for those components that have failed. For the components that did not fail before the test was terminated, all we know is that their failure times were longer than the duration of the test. As another example, in recording fire suppression times, the exact time of suppression may not be known; in some cases, all that may be available is an interval estimate (e.g., between 10 and 20 min). In this chapter, we will examine how to treat all of these cases, which we refer to generally as *uncertain data*.

10.1 Censored Data for Random Durations

Consider the following example, from [1]. A programmable logic controller (PLC) is being tested. Ten PLCs are placed in test, and each test is to be run for 1,000 h. If a PLC fails before the end of the test, its failure time is recorded. Assume that two of the PLCs failed during the test, at 395 and 982 h. The other eight PLCs were still operating when the test was terminated at 1,000 h (this is referred to in the literature as *Type I* censoring). How can we use this information to carry out Bayesian inference for the PLC failure rate, λ, assuming that the time to failure can be described by an exponential distribution?

D. Kelly and C. Smith, *Bayesian Inference for Probabilistic Risk Assessment*,
Springer Series in Reliability Engineering, DOI: 10.1007/978-1-84996-187-5_10,
© Springer-Verlag London Limited 2011

Table 10.1 OpenBUGS script for modeling censored random exponential durations.(Type I censoring)

```
model    {
  for(i in 1:N) {
  #Exponential likelihood function with Type I censoring
  t[i] ~ dexp(lambda)C(lower[i],)
  }
lambda ~ dgamma(0.0001, 0.0001)    #Jeffreys prior for lambda
}
data
list(t = c(982,394.7,NA,NA,NA,NA,NA,NA,NA,NA),
lower = c(982,394.7,1000,1000,1000,1000,1000,1000,1000,1000), N = 10)
Inits
list(lambda = 0.001)
list(lambda = 0.0001)
```

Frequentist estimation for censored data can be quite complicated, especially for nonexponential renewal distributions. The Bayesian approach implemented in OpenBUGS is very straightforward, and only slightly more complicated than the complete-sample cases considered in Chaps. 3 and 8. For more details on frequentist estimation, we refer the reader to [2, 3].

The key to dealing with censored times, for any likelihood distribution, is the C(lower, upper) construct in OpenBUGS. In cases where a failure time was not recorded, this instructs OpenBUGS to impute a failure time from the specified distribution, between the bounds specified by lower and upper in the construct. For our example above, lower would be equal to 1,000 h, and upper would be omitted, so we would have the following line in the OpenBUGS script: t[i] ~ dexp(lambda)C(1000,). With an exponential aleatory model, the likelihood function under Type I censoring is given by

$$ f(t_1, t_2, \cdots, t_r | \lambda) = \lambda^r \exp\left\{ -\lambda \left[\sum_{i=1}^{r} t_i + (n - r)t \right] \right\} $$

In this equation, we have observed failure times for r of the n components, leaving $n-r$ censored times, with t being the censoring time (i.e., the end of the observation period).

To incorporate censored data into OpenBUGS, a value of NA is entered in the data portion of the script for each censored failure time (time not recorded). Because lower is a vector, the failure time is entered if a failure time is observed, otherwise NA is entered. The script to analyze this example is shown in Table 10.1.

Running this script in the usual way gives a posterior mean for λ of 2.1×10^{-4}/h and a 90% credible interval of $(3.8 \times 10^{-5}, 5.1 \times 10^{-4})$.

Another way in which censoring could occur is if observation is terminated after a certain number of components have failed. Thus, failure times are not

Table 10.2 OpenBUGS script for Type II censoring of exponential failure times

```
model  {
for(i in 1:20)          {
          time[i]  ~  dexp(lambda)
          }
for(j in 21:N)  {
          time[j]  ~  dexp(lambda)C(time[20],)
          }
lambda  ~  dgamma(0.0001, 0.0001)
}
```

observed for the remaining components. In this case, the total observation period is random, and the censoring is referred to as *Type II censoring*. In this type of censoring, we observe the first r failure times, in order. The likelihood function for this type of censoring, assuming an exponential aleatory model, is given by:

$$f(t_{1:n}, t_{2:n}, \ldots, t_{r:n}) = \lambda^r \exp\left[-\sum_{i=1}^{r} t_{i:n} + (n-r)t_{r:n}\right]$$

Consider the following example of Type II censoring, from [2]. We have observed 30 components and recorded the first 20 times to failure, in days, as follows: 1, 3, 5, 7, 11, 11, 11, 12, 14, 14, 14, 16, 16, 20, 21, 23, 42, 47, 52, 62. Assuming these failures are exponentially distributed, we will find the posterior mean and 90% credible interval for the failure rate.

We have observed 20 failure times, but there were 30 components under observation. It is a mistake to use only the observed failure times to estimate λ, because we know that 10 of the components survived longer than 62 days, and this is substantial information about λ. The OpenBUGS script in Table 10.2 shows how to incorporate this information using the C (,) construct.

Running this script in the usual way we find a posterior mean for λ of 0.02/day and a 90% credible interval of (0.01, 0.03). Had we ignored the 10 components that survived longer than 62 days, that is, had we estimated λ treating the 20 observed failure times as a complete sample, we would have found a posterior mean of 0.05, with a 90% interval of (0.03, 0.07), which could be a significant over-estimate.

If the aleatory model is not exponential, as discussed in Chap. 8, frequentist estimation with censored data can become very difficult, and approximations often have to be employed to make the analysis tractable. However, with OpenBUGS, the analysis is straightforward using the C (,) construct. To illustrate this, consider the following example, taken from [1].

Consider the following data (in days) on time between failure: < 1, 5, < 10, 15, 4, 20, 30, 3, 30–60, 25. Let us find the probability that a component will operate longer than 20 days, assuming a Weibull distribution as the aleatory model for the failure time.

Table 10.3 OpenBUGS script for interval-censored failure times with Weibull likelihood

```
model
{
  for (i in 1 : N) {
  time.fail[i] ~ dweib(alpha, scale)C(lower[i], upper[i])#Weibull distribution for times between
  failures
  }
alpha ~ dgamma(0.0001,0.0001) #Diffuse priors for Weibull parameters
scale ~ dgamma(0.0001,0.0001)
prob.surv <- exp(-scale*pow(time.crit, alpha))#Probability of survival beyond time.crit
time.crit <- 20
}
data
list(time.fail = c(NA, 5, NA, 15, 4, NA, 3, NA, 25), N = 9)
list(lower = c(0, 5, 0, 15, 4, 20, 3, 30, 25), upper = c(1, 5, 10, 15, 4, 30, 3, 60, 25))
inits
list(alpha = 1, scale = 10) #initial values
list(alpha = 0.5, scale = 15)
```

Chapter 8 discussed Bayesian inference for the Weibull distribution. We mimic that approach here, but now our observed failure times are *interval-censored*. For example, we only know that the first component failed in less than a day, the third component in less than 10 days after the second, etc. We use the OpenBUGS script in Table 10.3 to analyze this example.

Running this script, and checking for convergence, we find the probability of survival for longer than 20 days has a posterior mean of 0.26 and a 90% credible interval of (0.09, 0.47).

10.2 Uncertainty in Binomial Demands or Poisson Exposure Time

In the analysis of the binomial and Poisson aleatory models in Chap. 3, we treated the number of demands and exposure time as fixed. However, in many practical problems encountered in PRA data analysis, there may actually be uncertainty associated with these parameters.

In the Bayesian approach, in which all uncertainties are represented via probability, the analyst specifies a distribution for the number of demands or the exposure time. This distribution represents the analyst's epistemic uncertainty as to the actual number of demands or exposure time. Often this might be a uniform distribution between known lower and upper bounds. The posterior distribution for the parameter of interest is then averaged over this distribution.

Consider the following example, taken from [4], in which we are interested in plugging of service water strainers. Assume we have two parallel strainers

Table 10.4 OpenBUGS script for estimating strainer-plugging rate when exposure time is uncertain

```
model      {
x ~ dpois(mu) #Poisson model for number of events, x
mu <- lambda*time.exp
lambda ~ dgamma(0.5, 0.0001) #Jeffreys prior for lambda
time.exp ~ dunif(24,140, 48,180) #Models uncertainty in exposure time
}
data
list(x = 4)
```

Table 10.5 OpenBUGS script script for estimating MOV failure probability when demands are uncertain

```
model      {
x ~ dbin(p, N)
N ~ dunif(275, 440) #Models uncertainty in demands
p ~ dbeta(0.5, 0.5) #Jeffreys prior for p, monitor this node
}
data
list(x = 4)
```

in a continually operating system, but we do not know if both strainers are always in service. Thus, over an observation time of 24,140 h, our estimate of exposure time, t, in a Poisson aleatory model could be as low as 24,140 and as high as 48,180 strainer-h.

The OpenBUGS script used for this example is shown in Table 10.4. Assume we have observed 4 strainer-plugging events during the uncertain exposure time, which we model as being uniformly distributed between the lower limit of 24,140 and the upper limit of 48,180 strainer-h. We use the Jeffreys prior distribution for λ. Running this script gives a posterior mean for λ of 1.3×10^{-4}/h and a 90% credible interval of $(4.5 \times 10^{-5}, 2.6 \times 10^{-4})$. Note that this is nearly what we would get by using the mean of our assumed exposure time distribution (36,160 strainer-h).

The same concept can be applied to the case of failure on demand, where the number of demands, n, in the binomial aleatory model may not be known accurately. Let us consider an example of motor-operated valves failing to open on demand, taken from [4]. Assume that the number of valve demands is nominally 381, but could have been as high as 440, and as low as 275. Again, let us assign n a uniform distribution over this range. We will use the Jeffreys prior for p. Assume that we have seen four failures of the valve to open.

The OpenBUGS script for this example is shown in Table 10.5. The posterior mean of p is estimated to be 0.01, the same value to 2 decimal places we would have obtained with the number of demands treated as known (381). Again, because our central estimate of n is near the fixed value of 381, modeling the uncertainty in n does not affect the posterior distribution of p very much. In many cases, we may find that modeling the uncertainty in n does not affect our estimate of p significantly.

As a final observation on these examples, even though we used a conjugate prior distribution in both examples, the posterior distributions are not gamma or beta, because we have two uncertain parameters instead of one.

10.3 Uncertainty in Binomial or Poisson Failure Counts

There are also occasions where the event or failure count is uncertain. For example, in examining equipment records, one may not be able to tell if a particular event constituted failure of equipment with respect to its function in the PRA model. Within the Bayesian framework, one can assign a subjective distribution to the event or failure counts, and the posterior distribution for the parameter of interest is then averaged over this distribution, giving the marginal posterior distribution for the parameter of interest. This is called the *posterior-averaging* approach in [5], and has been illustrated for Poisson event counts in [6, 7]. Alternative approaches to Bayesian inference with uncertain event or failure counts have been proposed, based on modifications to the likelihood function or treatment via imprecise probability. We will discuss several of these here. For more details, see [8, 9] for approaches based on modifying the likelihood function, and [10] for the approach based on imprecise probability.

Under the posterior-averaging approach, the posterior distribution for λ in the Poisson aleatory model becomes, considering uncertainty in both x and t:

$$g_{\text{avg}}(\lambda) = \sum_{i=1}^{N} \left[\int_{t_{\text{lower}}}^{t_{\text{upper}}} \frac{f(x_i|\lambda, t)g(\lambda)}{\int_0^\infty \int_{t_{\text{lower}}}^{t_{\text{upper}}} f(x_i|\lambda, t)g(\lambda)\pi(t)dtd\lambda} \pi(t)dt \right] \Pr(x_i)$$

The equation for the binomial case is similar.

To illustrate the approach, we revisit the strainer-plugging example from above. Earlier, the observed number of failures was treated as known (four). Now assume that plant records described seven plugging events over the time period of interest, but it was unclear if three of these events would have been considered as plugging events from the perspective of the PRA model. Therefore, the actual number of plugging events is uncertain, and could be four, five, six, or seven. In the posterior-averaging approach, the analyst assigns a subjective (discrete) distribution to these values, representing his confidence in the correctness of each value. Assume the analyst, in poring over the equipment records, has arrived at the following (discrete) distribution for the observed data:

$$\Pr(x = 4) = 0.75$$
$$\Pr(x = 5) = 0.15$$
$$\Pr(x = 6) = 0.075$$
$$\Pr(x = 7) = 0.025$$

Table 10.6 OpenBUGS script for estimating strainer-plugging by averaging posterior distribution over uncertainty in event count and exposure time

```
model     {
for(i in 1:N)          {
        x[i] ~ dpois(mu[i])
        mu[i] <- lambda[i]*time.exp
        lambda[i] ~ dgamma(0.5, 0.0001) #Jeffreys prior for lambda
        }
lambda.avg <- lambda[r] #Overall composite lambda, monitor this node
r ~ dcat(p[])
time.exp ~ dunif(24140, 48180) #Models uncertainty in exposure time
}
data
list(x = c(4,5,6,7), p = c(0.75, 0.15, 0.075, 0.025))
```

Note that these probabilities must sum to unity. OpenBUGS is used to analyze this problem numerically, via the script shown in Table 10.6. The `dcat()` distribution is used to model the discrete distribution for the failure count. This distribution has four components in this example, specified by the p vector in the data section of the script. OpenBUGS samples an integer between 1 and 4 from this distribution, using the probabilities listed above, which the analyst has developed, and uses the associated component of the failure count vector, x, to update the prior distribution for λ. Thus, OpenBUGS arrives at four posterior distributions for λ, which it then averages over the discrete distribution for the failure count to arrive at the posterior-average distribution for λ, monitored by the node `lambda.avg`. For the case where the exposure time is known with certainty to be 48,180 h, the posterior mean of this node is 1.0×10^{-4} h^{-1}, with 90% credible interval $(3.7 \times 10^{-5}$/h, 1.9×10^{-4}/h). If we include the uncertainty in the exposure time, we find the posterior mean to be 1.5×10^{-4}/h, with 90% credible interval $(5.2 \times 10^{-5}$/h, 3.0×10^{-4}/h).

As a second example, consider the earlier example of MOV failures under a binomial aleatory model. In this example, we took the observed number of MOV failures to be four. Consider now the case where this value is uncertain, and assume it could be 3, 4, 5, or 6, with Pr(3) = 0.1, Pr(4) = 0.7, Pr(5) = 0.15, and Pr(6) = 0.05. Using the OpenBUGS script shown in Table 10.7, for the case where the number of demands is known with certainty to be 381, we get a posterior mean for p node (`p.avg`) of 0.01, with 90% credible interval (0.004, 0.02). If the uncertainty in the number of demands is included as before, the posterior mean of p is still 0.01 to 2 decimal places, and the 90% credible interval has shifted slightly to (0.005, 0.03).

Table 10.7 OpenBUGS script for estimating MOV failure probability by averaging posterior distribution over uncertainty in failure count and number of demands

```
model    {
for (i in 1:N)          {
            x[i] ~ dbin(p[i], D)
            p[i] ~ dbeta(0.5, 0.5) #Jeffreys prior
            }
p.avg <- p[r] #Composite posterior, monitor this node
r ~ dcat(q[])
D ~ dunif(275, 440) #Models uncertainty in demands
}
data
list(x = c(3, 4, 5, 6), q = c(0.1, 0.7, 0.15, 0.05))
```

10.4 Alternative Approaches for Including Uncertainty in Poisson or Binomial Event Counts

We discuss two alternatives to posterior-averaging that have been proposed. We refer to these approaches as *likelihood-averaging* and *weighted likelihood*; both approaches incorporate the event count uncertainty into the likelihood function instead of averaging over multiple posterior distributions.

We first use a simple example to contrast the two likelihood-based approaches with posterior-averaging. In this example, we consider a single demand in which the outcome is either success ($X = 0$) with probability $1 - p$ or failure ($X = 1$) with probability p. For ease of illustration, we take the prior distribution for p to be uniform $(0, 1)$, which is a beta $(1, 1)$ distribution. Therefore, the posterior distribution of p will be a beta $(x + 1, 2 - x)$ distribution, since the beta distribution is conjugate to the Bernoulli aleatory model, as discussed in Chap. 3.

As an aside, note first that we could treat the uncertainty in x via interval-censoring, as we did above for random durations, using the C (,) construct in OpenBUGS. If we were to treat this example as being interval-censored, all we would know is that the value of X is either 0 or 1; in other words, we have gained no information from this demand, because 0 and 1 are the possible values of X. We would thus expect the posterior distribution to be unchanged from the prior distribution. The likelihood function is the probability that $X = 0$ plus the probability that $X = 1$, conditional upon a value of p. Thus, the likelihood function is

$$f(0|p) + f(1|p) = 1 - p + p = 1$$

The prior distribution for p is uniform, so the posterior is also uniform, as expected since a uniform prior is conjugate with the Bernoulli model. In fact, in this example the posterior will be the same as the prior for any prior distribution.

In the posterior-averaging approach, where we weight the two possible posteriors [$g(p|0)$ and $g(p|1)$] with weight w, we obtain the marginal posterior distribution for p by averaging over the possible values of X (0 and 1). The averaging yields

$$g(p) = w\,g(p|0) + (1 - w)g(p|1)$$

When the two values are judged equally likely ($w = 0.5$) and when $x = 0$, we have $g(p|0) = beta(1, 2) = 2(1 - p)$.

Alternatively, when $x = 1$, we have $g(p|1) = beta(2, 1) = 2p$.

Therefore, the marginal (weighted-average) posterior is given by

$$g(p) = \frac{2(1 - p) + 2p}{2} = 1$$

So the marginal posterior distribution is a uniform $(0, 1)$ distribution, just as we obtained in the interval-censored case. With unequal weights, the marginal posterior distribution in the posterior averaging approach will be given by

$$g(p) = \frac{w(1 - p) + (1 - w)p}{2} = \frac{w + (1 - 2w)p}{2}$$

In the *likelihood–averaging* approach, the likelihood function used to update the prior is a weighted average

$$\langle f(x|p) \rangle = wf(0|p) + (1 - w)f(1|p) = w + p(1 - 2w)$$

In the case where the values of x are judged to be equally likely ($w = 0.5$) the likelihood function reduces to 0.5. In this case, the posterior distribution is equal to the prior, which is uniform $(0, 1)$. If the values are not equally likely, then the posterior will be given by $g(p) = [w + p(1 - 2w)]/k$, where

$$k = \int_0^1 [w + (1 - 2w)p]dp = \frac{1}{2}$$

In the *weighted-likelihood* approach, the subjective event count probabilities developed by the analyst are used as exponential weighting factors for the likelihood function. In the case here of a single trial of the Bernoulli model, the result is

$$f(x|p) = \prod_{i=1}^2 [p^{x_i}(1 - p)^{1 - x_i}]^{w_i}$$

The effect of this likelihood function is as if there are $(w_i\, x_i)$ failures in $w_i\, (1 - x_i)$ demands. In the case of a single trial with a uniform prior on p, this gives a beta $(1.5, 1.5)$ posterior distribution. Thus, the posterior mean of p is 0.5, as with the other approaches, but the 90% credible interval is $(0.1, 0.9)$, narrower than in the other approaches. The posterior density for p obtained from the weighted-likelihood approach is shown in Fig. 10.1, and differs quite markedly from the uniform posterior density obtained in the other approaches.

Let us now consider the simple Bernoulli example, but with an informative prior for p. We will examine what happens when the prior distribution favors small values of p. Following [11], we will use a beta $(0.63, 123)$ distribution to illustrate

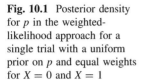

Fig. 10.1 Posterior density for p in the weighted-likelihood approach for a single trial with a uniform prior on p and equal weights for $X = 0$ and $X = 1$

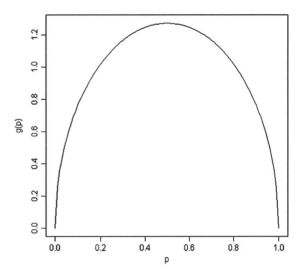

this case. The prior mean of p is 5.1E-3, with a 90% credible interval of (5.9E-5, 1.8E-2).

With equal weights on the possible outcomes of a single trial for the Bernoulli model (i.e., 0 or 1), the interval-censoring and likelihood-averaging approaches give the same outcome, which is just the prior distribution. In the interval-censoring approach, the probability of an outcome is either p or $1 - p$, so the weights are irrelevant. In the average-likelihood approach, the equal weights cancel in the numerator and denominator of Bayes' Theorem, giving the same effect. Thus, in these two approaches, the posterior mean of p is equal to the prior mean, which was 5.1E-3. The posterior-averaging approach gives a different result, because the weights are retained. The posterior mean of p increases from the prior mean of 5.1E-3 to 9.1E-3, and the 90% credible interval shifts to (1.7E-4, 2.8E-2). The weighted-likelihood approach gives a similar posterior mean, 9.0E-3, but a narrower 90% credible interval, (6.4E-4, 2.6E-2).

Because the equal weights cancel in the likelihood-averaging approach, if an analyst's reason for assigning equal weights is because he is expressing indifference to the two possible values of X, then the average-likelihood approach will reflect this indifference, while the posterior-averaging approach will not.

Now consider the case of unequal weights. Recall that the beta (0.63, 123) prior distribution weights small values of p heavily, and so the most likely outcome of a trial with this prior is $X = 0$ (probability of this outcome is 0.995). Imagine that one such trial is performed, and the actual outcome is not observed with complete certainty, but information is available such that the analyst judges that the outcome was $X = 1$ with probability 0.9. In this case, we would expect the posterior distribution for p to be close to what we would get by updating the prior with $X = 1$. The posterior-averaging approach behaves in this way, giving a posterior mean for

Table 10.8 OpenBUGS script for posterior-averaging approach to uncertainty in mixing valve failure count

```
model      {
for(i in 1:K) {
         x[i] ~ dbin(p[i], n) # Binomial distribution for number of failures
         p[i] ~ dlnorm(mu, tau) # Lognormal prior distribution for p
         }
p.avg <- p[r] # Monitor this node
r ~ dcat(p.analyst[])
for(j in 1:3) {
         p.analyst[j] <- 1/3
         }
# Calculate tau from lognormal error factor
tau <- pow(log(prior.EF)/1.645, -2)
# Calculate mu from lognormal prior median and error factor
mu <- log(prior.median)
}
data
list(x = c(1,2,3),n = 187, prior.median = 0.001, prior.EF = 5, K = 3)
```

p of 0.012, compared to the value of 0.013 we would have obtained had the outcome been $X = 1$ with certainty. The 90% credible interval for p is (1.1E-3, 3.2E-2). The outcome of the weighted-likelihood approach is similar, with a posterior mean of 0.012 and a somewhat narrower 90% credible interval of (1.5E-3, 3.1E-2). In contrast, the likelihood-averaging approach results in a posterior mean of 0.005; *the posterior distribution is essentially unchanged from the prior by the information from a single trial, even though the analyst was highly confident that the outcome was $X = 1$.* This aspect of the likelihood-averaging approach, its tendency to over-rule the data and allow the prior distribution to dominate the outcome, is a pitfall for using this approach with relatively sparse data.

We will summarize these three approaches (posterior-averaging, likelihood-averaging, and weighted likelihood) to incorporating uncertainty in event counts with an example taken from [11]. In this example we have a binomial aleatory model for failure of a mixing valve in an avionics thermal control system. The event of interest in the PRA model is failure of the mixing valve to change position in response to a signal from the avionics controller. Assume the prior distribution of the failure probability is lognormal with a median value of 0.001 and an error factor of 5. Records for the valve indicate there has been 1 failure in 187 demands. However, personnel responsible for maintaining the system indicate that there may have been two other failures, which were not recorded properly. Thus, the actual failure count is uncertain, with possible values 1, 2, or 3 (Tables 10.8, 10.9, 10.10).

We will first consider the case of equal weights for each of the possible failure counts. The OpenBUGS scripts for each approach are shown in

Table 10.9 OpenBUGS script for likelihood-averaging approach to uncertainty in mixing valve failure count

```
model    {
for(i in 1:K) {
      phi[i] <- w[i]*exp(logfact(n) - logfact(x[i]) - logfact(n-x[i]))*pow(p, x[i])*pow(1-p, n-x[i])
      w[i] <- 1/3
}
phi.sum <-sum(phi[])
log.phi.sum <-log(phi.sum)
zero <- 0
zero ~ dloglik(log.phi.sum)
p ~ dlnorm(mu, tau) #Lognormal prior distribution for p
# Calculate tau from lognormal error factor
tau <- pow(log(prior.EF)/1.645, -2)
# Calculate mu from lognormal prior median and error factor
mu <- log(prior.median)
}
data
list(x = c(1,2,3),n = 187, prior.median = 0.001, prior.EF = 5, K = 3)
```

Table 10.10 OpenBUGS script for weighted-likelihood approach to uncertainty in mixing valve failure count

```
model     {
for(i in 1:K){
         zeros[i] <- 0
         zeros[i] ~ dloglik(phi[i])#Phi is log-likelihood
         phi[i] <- w[i]*(logfact(n) - logfact(x[i]) - logfact(n-x[i]) + x[i]*log(p) + (n-x[i])*log
(1-p))
         w[i] <- 1/K
         }
p ~ dlnorm(mu, tau) #Lognormal prior distribution for p
# Calculate tau from lognormal error factor
tau <- pow(log(prior.EF)/1.645, -2)
# Calculate mu from lognormal prior median and error factor
mu <- log(prior.median)
}
data
list(x = c(1,2,3), n = 187, prior.median = 0.001, prior.EF = 5, K = 3)
```

Tables 10.9, 10.10, and the results are summarized in Table 10.11. As expected from the results for the simple Bernoulli example above, the likelihood-averaging approach with equal weights is equivalent to interval censoring (see Ex. 10.1), and the posterior-averaging and weighted-likelihood approaches give similar numerical results, with the weighted-likelihood approach capturing less of the uncertainty present in the problem.

Table 10.11 Summary of results for mixing valve example with equal weights

Approach	Posterior mean	90% Interval
Interval censoring	2.9E-3	4.6E-4, 8.1E-3
Posterior-averaging	4.4E-3	6.4E-4, 1.2E-2
Likelihood-averaging	2.8E-3	4.6E-4, 8.0E-3
Weighted likelihood	4.2E-3	8.7E-4, 1.0E-2

Table 10.12 Summary of results for mixing valve example with unequal weights

Approach	Posterior mean	90% Interval
Posterior-averaging	5.8E-3	1.0E-3, 1.4E-2
Likelihood-averaging	3.7E-3	5.3E-4, 1.E-2
Weighted likelihood	5.6E-3	1.3E-3, 1.3E-2

With unequal weights developed by the analyst, the interval-censoring approach is no longer applicable, as it does not use weights developed by the analyst. Let us assume in our mixing valve example that the analyst has developed the following subjective probabilities for the possible failure counts:

- $P(X = 1) = 0.1$
- $P(X = 2) = 0.2$
- $P(X = 3) = 0.7$

Note that the analyst has put a high probability on $X = 3$, which is a very unlikely outcome with the specified lognormal prior (probability $= 0.01$). We enter these weights and rerun the scripts above to obtain the results shown in Table 10.12. The likelihood-averaging results are somewhat nonconservative, and the weighted-likelihood results are what would be obtained by updating the lognormal prior with the weighted-average failure count (2.6) and 187 demands. When the possible outcomes are more consistent with the prior distribution, the posterior-averaging and likelihood-averaging approaches will generally be similar.

In summary, when binomial or Poisson event counts are uncertain, there are various approaches that can be taken to incorporating this uncertainty in the Bayesian model. In some limited cases, the outcomes of all the approaches will coincide; however in most cases they will not. The reader should keep in mind that other approaches are available, and there is as yet no consensus on a single approach. We summarize, in Table 10.13, some key aspects of each of the approaches.

10.5 Uncertainty in Common-Cause Event Counts

The multinomial distribution is the most commonly used aleatory model for common-cause failure (CCF) and its conjugate prior is the Dirichlet distribution. The observed data consist of a vector of failure counts, $\boldsymbol{n} = (n_1, n_2, ..., n_m)$, where

Table 10.13 Summary of approaches to Bayesian inference with uncertain binomial or Poisson event counts

Approach	Salient features
Interval censoring	Appropriate when data are known to lie in an interval, but information is not available to develop probability distribution for "true" event count
Posterior-averaging	Can lead to multimodal posterior
	Not equivalent to interval censoring when weights are equal
Likelihood-averaging	Equivalent to interval censoring when all weights are equal
	Weights effectively ignored when data are sparse, leading to domination by prior distribution
	Can lead to multimodal posterior
Weighted likelihood	Always leads to unimodal posterior
	Under-estimates uncertainty
	Equivalent to inference with weighted-average event count

n_k represents the number of events in which k redundant components failed in a common-cause component group (CCCG) of size m. The parameters of the multinomial distribution are commonly designated as α, the eponymous parameters of the alpha-factor method.

In typical analyses, the vector of observed failure counts is assumed to be known with certainty; in reality, judgment often must be used in developing $\mathbf{n}$. This uncertainty arises from two principal sources. The first source is that the actual impact on a plant experiencing a failure event may not be known. The second one is the uncertainty in mapping between CCCGs of different sizes. It is common to represent CCF data uncertainty in terms of an impact vector. For event j, the impact vector is

$$p_j = \left(p_{0,j}, p_{1,j}, \ldots, p_{k,j}\right)$$

where $p_{k,j}$ represents the probability that the event being analyzed involves the failure of k components in the CCCG; $p_{0,j}$ is the probability that the event is irrelevant to the CCCG being analyzed.

To illustrate this, consider the data for a three-pump system in Table 10.14, taken from [12]. If we assume that Events 3 and 5 arose from different causes, they can be treated as independent from a state-of-knowledge perspective. However, if Events 1, 2, and 8 all arose from the same cause, this state-of-knowledge dependence needs to be reflected in the uncertainty treatment.

We can use an event tree structure to develop the possible data sets, and then perform posterior-averaging using weights calculated from this tree. Figure 10.2 shows the event tree used in [12].

We can now use OpenBUGS to perform Bayesian inference for this data to estimate the distributions for the CCF alpha-factors. The script in Table 10.15 can be used to do this. We use a Dirichlet prior with all parameters equal to 1 in this script; this is a multidimensional analog of a uniform distribution, and thus is a noninformative prior for the alpha-factors. Note that 111 individual failures have been included, as in [12], each with impact vector (0, 1, 0, 0). Also note that because indexing in OpenBUGS starts with 1, alpha[4] will correspond to α_3.

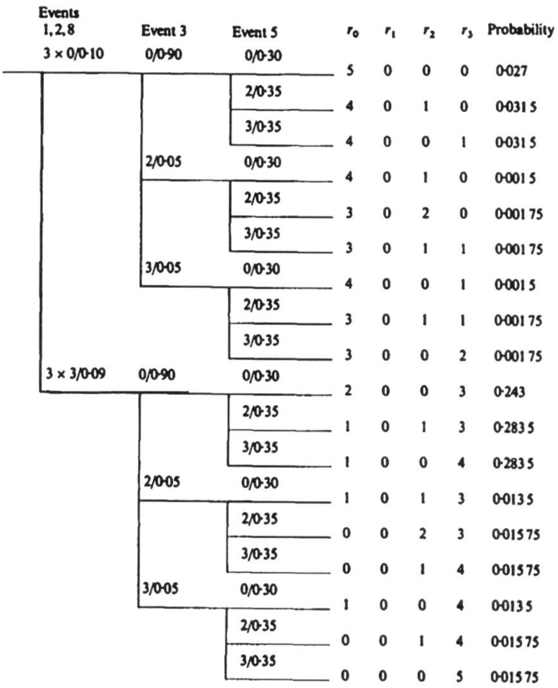

Fig. 10.2 Event tree for uncertain pump CCF data, taken from [12]

Events 1,2,8 3 × 0/0·10	Event 3 0/0·90	Event 5 0/0·30	r_0	r_1	r_2	r_3	Probability
			5	0	0	0	0·027
		2/0·35	4	0	1	0	0·0315
		3/0·35	4	0	0	1	0·0315
	2/0·05	0/0·30	4	0	1	0	0·0015
		2/0·35	3	0	2	0	0·00175
		3/0·35	3	0	1	1	0·00175
	3/0·05	0/0·30	4	0	0	1	0·0015
		2/0·35	3	0	1	1	0·00175
		3/0·35	3	0	0	2	0·00175
3 × 3/0·09	0/0·90	0/0·30	2	0	0	3	0·243
		2/0·35	1	0	1	3	0·2835
		3/0·35	1	0	0	4	0·2835
	2/0·05	0/0·30	1	0	1	3	0·0135
		2/0·35	0	0	2	3	0·01575
		3/0·35	0	0	1	4	0·01575
	3/0·05	0/0·30	1	0	0	4	0·0135
		2/0·35	0	0	1	4	0·01575
		3/0·35	0	0	0	5	0·01575

Table 10.14 Example CCF data for 3-pump CCCG, taken from [12]

Event	Impact vector
1	0.1, 0, 0, 0.9
2	0.1, 0, 0, 0.9
3	0.9, 0, 0.05, 0.05
4	0, 0, 1, 0
5	0.3, 0, 0.35, 0.35
6	0, 0, 1, 0
7	0, 1, 0, 0
8	0.1, 0, 0, 0.9

Running this script in the usual way we find the posterior means and 90% credible intervals for α shown in Table 10.16.

10.6   Exercises

1. For the mixing valve example, show that interval-censoring leads to the same posterior mean and 90% credible interval for p as the likelihood-averaging approach.

Table 10.15 OpenBUGS script for treating uncertainty in CCF event count

```
model     {
for(i in 1:M) {
          n[i,1:4] ~ dmulti(alpha[i,1:4], N[i]) #Multinomial distribution for observed event
counts
          n.rep[i,1:4] ~ dmulti(alpha[i,1:4], N[i]) #Posterior predictive failure count
          alpha[i, 1:4] ~ ddirch(theta[1:4]) #Dirichlet prior distribution for multinomial
parameters
#Note that alpha[4] is alpha-3 in the reference paper
          N[i] <- sum(n[i, 1:4])
          }
for(k in 1:K)         {
          alpha.avg[k] <- alpha[r,k] #Monitor this node for posterior-averaged values
          theta[k] <- 1 #Noninformative Dirichlet prior distribution
          }
#p[i] is the weight given to potential data vector i
r ~ dcat(p[])
}

data
list(K = 4, M = 18)
```

n[,1]	n[,2]	n[,3]	n[,4]	p[]
5	111	0	0	0.027
4	111	1	0	0.0315
4	111	0	1	0.0315
4	111	1	0	0.0015
3	111	2	0	0.00175
3	111	1	1	0.00175
4	111	0	1	0.0015
3	111	1	1	0.00175
3	111	0	2	0.00175
2	111	0	3	0.243
1	111	1	3	0.2835
1	111	0	4	0.2835
1	111	1	3	0.0135
0	111	2	3	0.01575
0	111	1	4	0.01575
1	111	0	4	0.0135
0	111	1	4	0.01575
0	111	0	5	0.01575

```
END
```

2. The following component repair times have been recorded, in hours: 3, 4.5, 2, 10, 5, 8.5. An additional four times have been recorded only approximately, to within the nearest hour: 3, 7, 11, 4. For each of these times, the actual repair time could have been 1 h less or 1 h greater than the recorded time. Treating these times as a random sample from an exponential model, and using

Table 10.16 Posterior distribution summaries for alpha-factors

	Mean	90% Credible interval
$\alpha_{1censoring}$	0.93	0.89, 0.97
α_2	0.01	6.8×10^{-4}, 0.03
α_3	0.03	0.007, 0.07

the Jeffreys prior, find the posterior mean and 90% credible interval for the exponential rate, λ.

3. The prior for the failure-to-close probability of a valve is lognormal with a median value of 0.0013 and a 95th percentile of 0.0087. Two failures of the valve have been observed, but the demand count is not certain. Assume there are three possible values for the demand count: 300, 400, 500. The probabilities that each of these is the true demand count have been assessed to be 0.2, 0.6, and 0.2, respectively. Assuming a binomial aleatory model for the number of failures to close, find the posterior mean and 90% credible interval for the failure-to-close probability.

4. Show that the weighted-likelihood approach with a conjugate prior is equivalent to updating the conjugate prior with weighted-average data.

5. Suppose that historical data for sodium valve failures suggest a uniform distribution for the valve failure rate, with bounds of 1.4×10^{-5}/h and 5.0×10^{-5}/h. Twenty such valves are placed on test and the test is run, without replacement, until five valves have failed. Assume that the fifth failure occurs at 160,000 h. Find the posterior mean and 95% credible interval for the valve failure rate.

References

1. Rodionov A, Kelly D, Uwe-Klügel J (2009) Guidelines for analysis of data related to ageing of nuclear power plant components and systems. Joint Research Centre, Institute for Energy, Luxembourg, European Commission
2. Bain L, Engelhardt M (1991) Statistical theory of reliability and life-testing models. Marcel Dekker, New York (Basel)
3. Meeker WO, Escobar LA (1998) Statistical methods for reliability data. Wiley, New York
4. Kelly DL, Smith CL (2009) Bayesian inference in probabilistic risk assessment: the current state of the art. Reliab Eng Syst Saf 94:628–643
5. Siu NO, Kelly DL (1998) Bayesian parameter estimation in probabilistic risk assessment. Reliab Eng Syst Saf 62:89–116
6. Martz H, Picard R (1995) Uncertainty in poisson event counts and exposure time in rate estimation. Reliab Eng Syst Saf 48:181–193
7. Martz H, Hamada M (2003) Uncertainty in counts and operating times in estimating poisson occurrence rates. Reliab Eng Syst Saf 80:75–79
8. Tan Z, Xi W (2003) Bayesian analysis with consideration of data uncertainty in a specific scenario. Reliab Eng Syst Saf 79:17–31
9. Groen FG, Mosleh A (2005) Foundations of probabilistic inference with uncertain evidence. Int J Approximate Reasoning 39:49–83

10. Walley P (1991) Statistical reasoning with imprecise probabilities. Chapman and Hall, London
11. Dezfuli H, Kelly DL, Smith C, Vedros K, Galyean W (2009) Bayesian inference for NASA probabilistic risk and reliability analysis. NASA, Washington, DC
12. Siu NO (1990) A monte carlo method for multiple parameter estimation in the presence of uncertain data. Reliab Eng Syst Saf 28:59–98

Chapter 11
Bayesian Regression Models

Sometimes a parameter in an aleatory model, such as p in the binomial distribution or λ in the Poisson distribution, can be affected by observable quantities such as pressure, mass, or temperature. For example, in the case of a pressure vessel, very high pressure and high temperature may be leading indicators of failures. In such cases, information about the explanatory variables can be used in the Bayesian inference paradigm to inform the estimates of p or λ. We have already seen examples of this in Chap. 5, where we modeled the influence of time on p and λ via logistic and loglinear regression models, respectively. In this chapter, we extend this concept to more complex situations, and illustrate the possibility of a Bayesian regression approach with several examples, the first of which estimates the probability of O-ring failure in the solid-rocket booster motors of the space shuttle, taken from [1].

11.1 Aleatory Model for O-ring Distress

Table 11.1 shows data on O-ring thermal stress collected during launches prior to the 1986 launch of the Challenger that led to disastrous failure of the O-rings.[1] Each shuttle has three primary and three secondary O-rings, with one of each type having to fail to cause a disaster such as the Challenger.

There are six O-rings on the shuttle, so during each launch, the number of distress events, defined as erosion or blow-by of a primary field O-ring, is modeled as binomial with parameters p and $n = 6$: $X \sim \text{binomial}(p, 6)$. In this model, p is a function of both temperature and applied leak-test pressure. The canonical link

---

[1] As pointed out by [2], inclusion of flights with no incidents of blowholes is questionable. Therefore, the analysis presented here is intended only to illustrate the types of modeling that are possible, and the results should not be interpreted as the output of a validated data set.

D. Kelly and C. Smith, *Bayesian Inference for Probabilistic Risk Assessment*,
Springer Series in Reliability Engineering, DOI: 10.1007/978-1-84996-187-5_11,
© Springer-Verlag London Limited 2011

Table 11.1 O-ring thermal stress data prior to launch of Challenger in January 1986.[a]

Flight	Distress[a]	Temp (°F)	Press (psig)
1	0	66	50
2	1	70	50
3	0	69	50
5	0	68	50
6	0	67	50
7	0	72	50
8	0	73	100
9	0	70	100
41-B	1	57	200
41-C	1	63	200
41-D	1	70	200
41-G	0	78	200
51-A	0	67	200
51-C	2	53	200
51-D	0	67	200
51-B	0	75	200
51-G	0	70	200
51-F	0	81	200
51-I	0	76	200
51-J	0	79	200
61-A	2	75	200
61-B	0	76	200
61-C	1	58	200

[a] Thermal distress is defined to be erosion of the O-ring or blow-by of hot gases. The table shows the number of distress events for each launch. There are six field O-rings on the shuttle, so the number of distress events is an integer in the interval [0, 6]

function is the logit function, which we encountered in Chap. 5 in the time-trend model for p:

$$\log it(p) = \ln\left(\frac{p}{1-p}\right)$$

Engineering knowledge leads us to believe that p will vary as pressure and temperature vary (i.e., the leading indicators mentioned earlier), so we construct a model for p with pressure and temperature as explanatory variables. We consider two potential explanatory models:

(1) $logit(p) = a + bT + cP$
(2) $logit(p) = a + bT$

The OpenBUGS script for the first model, which includes both temperature and pressure as explanatory variables, is shown in Table 11.2. Diffuse normal priors were used for the coefficients in this model to allow the numerical results to be compared with the maximum likelihood estimates and confidence intervals

Table 11.2 OpenBUGS script for logistic regression model of O-ring distress probability with pressure and temperature as explanatory variables

```
model    {
for(i in 1:K) {
          distress[i] ~ dbin(p[i], 6)
          # Regression model by Dalal et al with temp and pressure
          logit(p[i]) <- a + b*temp[i] + c*press[i]
          distress.rep[i] ~ dbin(p[i], 6) # Replicate values for model validation
          diff.obs[i] <- pow(distress[i] - 6*p[i], 2)/(6*p[i]*(1-p[i]))
          diff.rep[i] <- pow(distress.rep[i] - 6*p[i], 2)/(6*p[i]*(1-p[i]))
          }
chisq.obs <- sum(diff.obs[])
chisq.rep <- sum(diff.rep[])
p.value <- step(chisq.rep - chisq.obs)
distress.31 ~ dbin(p.31, 6) # Predicted number of distress events for launch 61-L
logit(p.31) <- a + b*31 + c*200                # Regression model with temp and pressure
                                               # from day of the accident

# Prior distributions
a ~ dnorm(0, 0.000001)
b ~ dnorm(0, 0.000001)
c ~ dnorm(0, 0.000001)
}
data
list(
  distress=c(0,1,0,0,0,0,0,0,1,1,1,0,0,2,0,0,0,0,0,0,2,0,1),
  temp=c(66,70,69,68,67,72,73,70,57,63,70,78,67,53,67,75,70,81,76,79,75,76,58),
  press=c(50,50,50,50,50,50,100,100,200,200,200,200,200,200,200,200,200,200,200,200,
    200,200),
  K=23
    )

inits
list(a=5, b=0, c=0)
list(a=1, b=-0.1, c=0.1)
```

Table 11.3 Summary posterior estimates of logistic regression parameters, with both temperature and pressure included as explanatory variables

Parameter	Mean	Standard Dev.	95% Interval
a (intercept)	2.24	3.74	(-4.71, 9.92)
b (temp. coeff.)	-0.105	0.05	(-0.20, -0.02)
c (press. coeff.)	0.01	0.009	(-0.004, 0.03)

Table 11.4 OpenBUGS script for logistic regression of primary O-ring distress on temperature only

```
model   {
for(i in 1:K)   {
            distress[i]  ~ dbin(p[i], 6)
            logit(p[i])  <- a + b*temp[i] #Model with temperature only
            }
distress.31  ~ dbin(p.31, 6)
logit(p.31)  <- a + b*31
a  ~ dflat() #Diffuse priors over real axis
b  ~ dflat()
}

Inits
list(a=1, b=0.1) #Chain 1
list(a=10, b=-0.1) #Chain 2
```

Table 11.5 Summary posterior estimates of logistic regression parameters, temperature included as explanatory variable

Parameter	Mean	Standard Dev.	95% Interval
a (intercept)	5.225	3.16	(-1.00, 11.48)
b (temp. coeff.)	-0.12	0.049	(-0.22, -0.025)

obtained by [3]. This script also predicts the probability of O-ring distress at the Challenger launch temperature of $31°F$ and pressure of 200 psig, information that presumably would have been of great value in the Challenger launch decision.

Five thousand burn-in iterations were used, followed by 100,000 iterations to estimate the parameters. Table 11.3 shows the posterior mean, standard deviation, and 95% credible interval for each of the parameters in the logistic regression model for p. The marginal posterior distributions for a and b are approximately normal with the listed posterior means and standard deviations. The expected number of distress events at $31°F$ is about four (indicating the distress of four of six O-rings) and the probability that one of the six O-rings experiences thermal stress during a launch at that temperature is about 0.71.

We next consider a simpler model, in which only temperature is included as an explanatory variable for the logistic regression. The WinBUGS script for this model is shown in Table 11.4.

One thousand burn-in iterations were required for convergence, followed by 100,000 iterations to estimate the parameters. Table 11.5 shows the posterior mean, standard deviation, and symmetric 95% interval for each of the parameters in the logistic regression model for p.

Reference [3] also examined a model that is quadratic in temperature. Specifically, they analyzed the following model for distress probability, where $\bar{t}$ is the average of the temperature readings in the data:

Table 11.6 OpenBUGS script for logistic regression of primary O-ring distress quadratic in temperature

```
model    {
for(i in 1:K)        {
   distress[i] ~ dbin(p[i], 6)
   logit(p[i]) <- a +b*(temp[i] - temp.mean) + c*pow(temp[i] - temp.mean, 2)
   }
temp.mean <- mean(temp[])
a ~ dnorm(0, 0.000001)
b ~ dnorm(0, 0.000001)
c~ dnorm(0, 0.000001)
}

Inits
list(a=-3, b=0.05, c=0.005)
list(a=0, b=-0.05, c=-0.005)
```

Table 11.7 Summary posterior estimates for logistic regression model for primary O-ring distress, quadratic model in temperature

Parameter	Mean	Standard Dev.	95% Interval
a (intercept)	-3.25	0.52	(-4.37, -2.32)
b (linear. coeff.)	-0.10	0.08	(-0.27, 0.03)
c (quad.. coeff.)	0.003	0.006	(-0.01, 0.01)

$$\log it(p) = a + b(t - \bar{t}) + c(t - \bar{t})^2$$

The OpenBUGS script for this model is shown in Table 11.6. We analyzed this model with 1,000 burn-in iterations, followed by 50,000 iterations for parameter estimation. The posterior distributions of the parameters are summarized in Table 11.7.

11.2 Model-Checking

We will use the Bayesian chi-square statistic discussed in Chap. 4 to calculate a Bayesian p-value as a quantitative measure of the predictive model validity. We will also use the deviance information criterion (DIC), discussed in Chap. 8, as a means of selecting among models based on their relative predictive ability. The excerpt of the OpenBUGS script used to calculate the Bayesian p-value is shown in Table 11.8.

The model with only temperature predicts essentially the same number of distress events as the two more complex models. The DIC is nearly the same for all the models; the simplest model with only temperature as an explanatory variable

Table 11.8 Portion of OpenBUGS script for calculating Bayesian p-value for regression models

```
model   {
for(i in 1:K)         {
distress[i] ~ dbin(p[i], 6)
logit(p[i]) <- a + b*temp[i] + c*press[i] #Model with temperature and pressure
            distress.rep[i] ~ dbin(p[i], 6) #Replicate from posterior predictive distribution
            diff.obs[i] <- pow(distress[i] - 6*p[i], 2)/(6*p[i]*(1-p[i]))
            diff.rep[i] <- pow(distress.rep[i] - 6*p[i], 2)/(6*p[i]*(1-p[i]))
            }
chisq.obs <- sum(diff.obs[]) #Observed summary statistic
chisq.rep <- sum(diff.rep[]) #Replicated summary statistic
p.value <- step(chisq.rep - chisq.obs) #Mean of this node should be near 0.5
}
```

Table 11.9 Model-checking results for logistic regression models for primary O-ring distress

Explanatory Variables	DIC	Bayesian p-value
Temperature and pressure	36.58	0.19
Temperature	35.75	0.21
Quadratic in temperature	37.18	0.20

has a slightly larger Bayesian p-value than the model with both temperature and pressure, and is essentially the same as the model that is quadratic in temperature. Because the simplest model is essentially equivalent to the more complex ones, we would recommend it for predictive analyses (Table 11.9).

11.3 Probability of Shuttle Failure

The probability of shuttle failure is given by the joint probability of (a) primary O-ring erosion, (b) primary O-ring blowby, (c) secondary O-ring erosion, and (d) secondary O-ring failure. The above analysis examined the probability of $a \cup b$, because we defined O-ring distress as either erosion or blowby. Following [3], we use p_a, p_b, p_c, and p_d to denote the probabilities of (a)–(d), conditional upon the preceding events. The probability of failure of a field joint is then given by the product of these conditional probabilities:

$$p_F = p_a p_b p_c p_d \qquad (11.1)$$

There are six field joints on each shuttle. Assuming the joint failures are independent and identically distributed, then the probability of shuttle failure due to field joint failure is the probability that at least one of the six joints fails:

$$p_{sh} = 1 - (1 - p_F)^6$$

Table 11.10 Data on primary O-ring erosion for shuttle field joints

Flight	Erosion	Temp (°F)	Press (psig)
1	0	66	50
2	1	70	50
3	0	69	50
5	0	68	50
6	0	67	50
7	0	72	50
8	0	73	100
9	0	70	100
41-B	1	57	200
41-C	1	63	200
41-D	1	70	200
41-G	0	78	200
51-A	0	67	200
51-C	2	53	200
51-D	0	67	200
51-B	0	75	200
51-G	0	70	200
51-F	0	81	200
51-I	0	76	200
51-J	0	79	200
61-A	0	75	200
61-B	0	76	200
61-C	1	58	200

Note that this assumption of independence is highly questionable; joint failures are likely to be dependent, and thus we are likely calculating a lower bound on the shuttle failure probability.

We now turn to estimating each of the inputs to Eq. 11.1. For p_a, we have the primary O-ring erosion data given in Table 11.10.

Reference [3] fit a logistic regression model for p_a using this data, with temperature and leak-test pressure as explanatory variables. They concluded that pressure was not a significant variable, but kept it in the model because NASA engineers had thought that it would be an important predictor of erosion. We performed a Bayesian analysis of this model, as above.

Table 11.11 shows the posterior mean, standard deviation, and 90% credible interval for each of the parameters in the logistic regression model for p. These results compare well with the MLEs and 90% bootstrap confidence intervals obtained by [3]. Zero is near the center of the marginal posterior distribution for the coefficient of pressure, indicating that pressure is not a significant explanatory variable, as was concluded by [3]. Note also the larger Bayesian p-value for the model with temperature alone, and the slightly smaller DIC, shown in Table 11.12. Together, these suggest that the model without pressure as an explanatory variable is better able to replicate the observed data. We will estimate

Table 11.11 Summary posterior estimates of logistic regression parameters for primary O-ring erosion, temperature and pressure included as explanatory variables

Parameter	Mean	Standard Dev.	90% Interval
a (intercept)	8.38	5.38	(0.43, 17.46)
b (temp. coeff.)	-0.19	0.07	(-0.31, -0.08)
c (press. coeff.)	0.004	0.01	(-0.01, 0.02)

Table 11.12 Model-checking results for logistic regression models for primary O-ring erosion

Explanatory Variables	DIC	Bayesian p-value
Temperature and pressure	26.84	0.34
Temperature	24.91	0.50

the probability of shuttle failure both with pressure included, and with the simpler model that only includes temperature as an explanatory variable.

We next consider the conditional probability of primary O-ring blowby, given erosion, p_b. Of the seven field joints that exhibited erosion, only two also exhibited blow-by. This is too sparse a sample for regression modeling, so we follow [3] and estimate p_b by pooling the data from field O-rings with data from nozzle O-rings, which exhibited similar performance (5 blow-by events in 17 erosion events). This gives a total of 7 blow-by events in 24 erosion events with which to estimate p_b. Reference [3] chose a uniform(0, 1) prior for p_b; the Jeffreys prior, which is a beta(0.5, 0.5) distribution, would be a more standard choice in PRA. However, the extra bias introduced by the uniform prior is minimal in this case, so we will retain the uniform prior for p_b.

We turn next to the conditional probability of secondary O-ring erosion, given blow-by of the primary O-ring, p_c. Again, there was very little data with which to quantify p_c. Reference [3] pooled data from field and nozzle O-rings, yielding two events out of seven in which primary O-ring blow-by led to erosion of the secondary O-ring. Again, [3] used a uniform prior for p_c.

No events existed in which a secondary O-ring had failed following primary O-ring erosion and blow-by, followed by secondary O-ring erosion. Therefore, [3] set p_d equal to p_b. This correlates the state of knowledge of p_d with that of p_b, and changes Eq. 11.1, giving the probability of field joint failure, to

$$p_F = p_a p_b^2 p_c \tag{11.2}$$

Table 11.13 shows the complete OpenBUGS script used to estimate each of the terms in Eq. 11.2 and to propagate the uncertainties represented by the posterior distributions of the conditional probabilities in this equation to obtain the distribution for shuttle failure probability as a result of O-ring failure.

Table 11.13 OpenBUGS script used to calculate probability of shuttle failure as a result of field joint failure

```
model    {
for(i in 1:K)        {
  erosion.prim[i] ~ dbin(p.a[i], 6)
  logit(p.a[i]) <- a + b*temp[i] + c*press[i]
  }
blowby.erode ~ dbin(p.b, n.erode.blby) #Binomial dist. for primary blowby, given erosion
n.erode.blby <- 24 #Pooled field and nozzle O-ring data
    p.b ~ dunif(0, 1) #Prior used by Dalal et al.
erode.sec ~ dbin(p.c, n.erode.sec) #Binomial dist. for sec. erosion, given primary erosion and
blowby
n.erode.sec <- 7
p.c ~ dunif(0, 1)
p.F.31 <- p.a.31*pow(p.b, 2)*p.c #Probability of field joint failure at 31 deg. F
p.F.60 <- p.a.60*pow(p.b, 2)*p.c #Probability of field joint failure at 60 deg. F
p.sh.31 <- 1 - pow(1-p.F.31, 6) #Probability of shuttle failure at 31 deg. F
p.sh.60 <- 1 - pow(1-p.F.60, 6) #Probability of shuttle failure at 60 deg. F
logit(p.a.31) <- a + b*31 + c*200
logit(p.a.60) <- a + b*60 + c*200
a ~ dnorm(0, 0.000001) #Diffuse priors on logistic regression coefficients
b ~ dnorm(0, 0.000001)
c ~ dnorm(0, 0.000001)
}

data
list(blowby.erode=7, erode.sec=2)

inits
list(a=5, b=0, c=0) #Logistic model for temp and press
list(a=1, b=-0.1, c=0.1)
```

Table 11.14 Summary of posterior distributions for shuttle failure due to field joint failure

Explanatory Variables	Mean		90% Interval	
	31°F	60°F	31°F	60°F
Temperature and pressure	0.163	0.02	(0.03, 0.39)	(0.0035, 0.07)
Temperature	0.165	0.02	(0.03, 0.39)	(0.0035, 0.07)

This script was run with 1,000 burn-in samples, followed by 100,000 samples for parameter estimation. The results for shuttle failure probability are summarized in Table 11.14. As can be seen, the probability of shuttle failure is essentially independent of any effect due to leak-test pressure, and is significantly higher at the lower of the two temperatures. The posterior distributions for shuttle failure probability at the two different temperatures are shown in Fig. 11.1.

Fig. 11.1 Posterior
distribution for shuttle failure
probability at 31 and 60°F

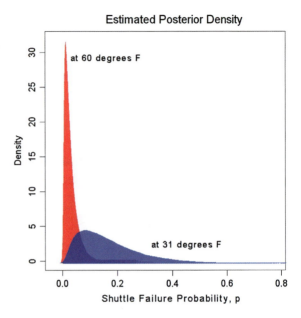

11.4 Incorporating Uncertainty in Launch Temperature

In the Bayesian framework, it is straightforward to incorporate uncertainty about the launch temperature to represent the decision-maker's state of knowledge in advance of the launch. This is a distinct advantage over the frequentist framework. To illustrate this concept, we will treat the temperature at launch as uncertain, as it would be during the launch planning stages. We will assume the average launch temperature in January is equal to the average low temperature (from www.weatherbase.com) of 52°F. We will further assume that the lowest reported temperature during January (26°F) represents a difference of two standard deviations. We will thus take the predicted launch temperature as being normally distributed with a mean of 52°F and a standard deviation of 13°F.

This requires only a slight modification to the OpenBUGS script. Table 11.15 shows the revised script. The predicted mean shuttle failure probability due to O-ring failure is now 0.08, with a 90% credible interval of (0.001, 0.27). This illustrates the value of knowing the launch temperature; it refines the estimate of shuttle failure probability significantly, providing more information to the decision-maker.

11.5 Regression Models for Component Lifetime

The preceding example focused on count data. Regression models are also useful for incorporating physical variables such as temperature into aleatory models for component failure time. In the reliability community, it is common to test highly

Table 11.15 OpenBUGS script for predicting shuttle failure probability when launch temperature is uncertain

```
model    {
for(i in 1:K)       {
            erosion.prim[i]  ~  dbin(p.a[i], 6)
logit(p.a[i])  <- a + b*temp[i] + c*press[i] #Model with temperature and pressure
            erosion.prim.rep[i]  ~  dbin(p.a[i], 6)
            }
blowby.erode  ~  dbin(p.b, n.erode.blby) #Binomial dist. for primary blowby, given erosion
n.erode.blby  <- 24 #Pooled field and nozzle O-ring data
p.b  ~  dunif(0, 1) #Prior used by Dalal et al.
erode.sec  ~  dbin(p.c, n.erode.sec) #Binomial dist. for sec. erosion, given primary erosion and
blowby
n.erode.sec  <- 7
p.c  ~  dunif(0, 1)
p.F.pred  <- p.a.pred*pow(p.b, 2)*p.c
p.sh.pred  <- 1 - pow(1-p.F.pred, 6)
erosion.prim.pred  ~  dbin(p.a.pred, 6)
logit(p.a.pred)  <- a + b*temp.pred + c*200
temp.pred  ~  dnorm(52, 0.006)
a  ~  dnorm(0, 0.000001)
b  ~  dnorm(0, 0.000001)
c  ~  dnorm(0, 0.000001)
}
```

reliable components under conditions that are more severe than the environment in which the components will be operated. Such accelerated life testing (ALT) is performed to provide more abundant failure data with which to estimate a component failure rate than could be obtained under normal operating conditions. However, to provide a failure rate estimate that is valid under normal conditions, the effects of the stressors introduced in the ALT must be taken into account. A common model for doing this is the Arrhenius model, first proposed as a model for temperature dependence of chemical reaction rates.

In the Arrhenius model, the rate of interest, r, is a function of temperature, T, according to

$$r(T) = C \exp\left(-\frac{E_A}{kT}\right)$$

In this equation, C and E_A are specific to the material being tested, k is Boltzmann's constant, and T is the temperature on an absolute scale (e.g., Kelvin). We illustrate the application of the Arrhenius model to component lifetime data with the following example. Assume that we are observing time to failure of a component for which higher temperatures tend to lead to earlier failure times, that

Table 11.16 Failure times of component at four different temperatures, from [4]

300K	350K	400K	500K
100^*	47.5	29.5	80.9
100^*	73.7	100^*	76.6
100^*	100^*	52.0	53.4
80.7	100^*	63.5	100^*
100^*	86.2	100^*	47.5
29.1	100^*	99.5	26.1
100^*	100^*	56.3	77.6
100^*	100^*	92.5	100^*
100^*	100^*	100^*	61.8
100^*	71.8	100^*	56.1

* It indicates the component was still operating at 100 hr, when the test ended

is, high temperature is a stressor for the component. We will adopt a Weibull aleatory model for time to failure, parameterized as follows:

$$f(t|\beta, \psi) = \frac{\beta}{\psi} \left(\frac{t}{\psi}\right)^{\beta-1} \exp\left[-\left(\frac{t}{\psi}\right)^{\beta}\right]$$

This is related to the Weibull(β, λ) parameterization used in Chap. 8 by $\lambda = \psi^{-\beta}$. In this parameterization, the expected lifetime, conditional upon β and ψ, is given by $\psi\Gamma(1 + 1/\beta)$, where $\Gamma()$ is the gamma function. To incorporate the dependence upon temperature, we place an Arrhenius model on the Weibull scale parameter, ψ, which is the characteristic lifetime of a component:

$$\psi(T) = C\exp\left(-\frac{E_A}{kT}\right)$$

As temperature increases, ψ decreases, and thus the expected lifetime, which is proportional to ψ, also decreases. For Bayesian inference, it will be convenient to work with the logarithm of ψ. Reference [4] rewrite ψ as follows:

$$ln(\psi) = \alpha_0 + \frac{\alpha_1}{T}$$

where $\alpha_0 = -\ln(C)$ and $\alpha_1 = -E_A/k$.

As an example, consider the failure times listed in Table 11.16, from [4]. These are times collected at four different temperatures, under the assumption of a renewal process. Times with an asterisk indicate that the component had not yet failed. In the terminology of Chap. 10, these are censored failure times.

The OpenBUGS script in Table 11.17 is used to analyze these times. We have used improper flat priors on α_0 and α_1, and a diffuse uniform distribution on the Weibull shape parameter, β. We have generated a new time from the posterior predictive distribution, monitored by the node `time.pred`. This node will give the predicted lifetime of a component exposed to a temperature T_{pred}, which has been set to 293K in this script.

Table 11.17 OpenBUGS script for analyzing accelerated lifetime data in Table 11.16

```
model      {
for(i in 1:K)          {
for(j in 1:N)          {
                       time[i,j] ~ dweib(beta, lambda[i])C(lower[i,j],)
                       }
           lambda[i] <- pow(psi[i], -beta)
#Arrhenius model
           log.psi[i] <-alpha.0 + alpha.1/T[i]
       }
#Predictive distribution for component operating at T.pred
time.pred ~ dweib(beta, lambda.pred)
lambda.pred <- pow(psi.pred, -beta)
#Arrhenius model
log.psi.pred <- alpha.0 + alpha.1/T.pred
beta ~ dunif(0, 10)
alpha.0 ~ dflat()
alpha.1 ~ dflat()
}

data
list(T=c(300, 350, 400, 500), time=structure(
.Data=c(
NA,NA,NA,80.7,NA,29.1,NA,NA,NA,NA,
47.5,73.7,NA,NA,86.2,NA,NA,NA,NA,71.8,
29.5,NA,52.0,63.5,NA,99.5,56.3,92.5,NA,NA,
80.9,76.6,53.4,NA,47.5,26.1,77.6,NA,61.8,56.1),
.Dim=c(4, 10)),
lower=structure(
.Data=c(100,100,100,0,100,0,100,100,100,100,
0,0,100,100,0,100,100,100,100,0,
0,100,0,0,100,0,0,0,100,100,
0,0,0,100,0,0,0,100,0,0),
.Dim=c(4,10)),
   N=10, K=4, T.pred=293)

inits
Arrhenius model
list(beta=1, alpha.0=1, alpha.1=100)
list(beta=2, alpha.0=0, alpha.1=200)
list(beta=1, alpha.0=5, alpha.1=100)
list(beta=2, alpha.0=5, alpha.1=500)
```

Because of the relatively large number of parameters in this script, we use four chains to aid in checking for convergence. Convergence appears to be achieved within the first 10,000 samples, so we discard the first 10,000 samples for burn-in, and run an additional 100,000 iterations for parameter estimation. The resulting posterior means and 90% credible intervals are shown in Table 11.18.

Table 11.18 Results of Bayesian inference using OpenBUGS script in Table 11.17

	Mean	90% Interval
β	2.4	(1.6, 3.2)
α_0	3.0	(2.0, 4.0)
α_1	666.9	(288.3, 1111.0)
Predicted life	190.7	(52.6, 405.0)

11.6 Battery Example

To close this chapter, we present a comprehensive example that illustrates techniques developed in previous chapters. It also illustrates some common pitfalls associated with naively "turning the Bayesian crank," and shows how qualitative and quantitative checks can reveal problems with the model.

Assume we need to estimate battery reliability for a space mission time of 10,000 h. Assume that the decision-maker is interested in the mean and fifth percentile of battery reliability. Recall the general steps associated with Bayesian inference, given in Chap. 1:

1. Begin with an aleatory model for the process being represented in the PRA (e.g., failure of component to change state on demand)
2. Specify a prior distribution for parameter(s) in this model, quantifying epistemic uncertainty, that is, quantifying a state of knowledge about the possible parameter values
3. Observe data from or related to the process being represented
4. Update the prior to obtain the posterior distribution for the parameter(s) of interest
5. Check validity of the model, data, and prior.

We thus begin by specifying an aleatory model. Our observed data will be in the form of battery failure times, assumed to be a random sample from the aleatory model. As discussed in Chap. 3, the simplest aleatory model for such data is an exponential distribution, with unknown parameter λ. We begin with this simplest model.

The second step is to specify a prior distribution for λ, representing our state of knowledge about possible values of λ based on past information. For this example, assume the battery vendor has provided information in the form of a mean time to failure (MTTF) of 50,000 h, and has also provided a "lower bound" failure time of 40,000 h. The analyst decides to encode this information into a conjugate gamma prior distribution, taking 1/MTTF as the prior mean of λ, and the reciprocal of the "lower bound" as the 95th percentile of λ.

We can use a spreadsheet to find the parameters, α and β, of the gamma distribution encoding this information, as described in Chap. 3. We find the resulting prior is a gamma($52.2, 2.6 \times 10^6$ h) distribution, shown in Fig. 11.2.

Fig. 11.2 Gamma prior
distribution for battery failure
rate, λ

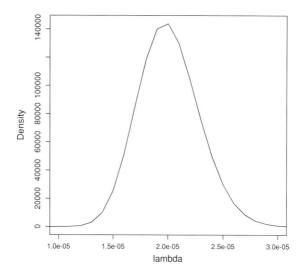

Step 3 in the Bayesian inference process is to collect battery failure times. We observe the following 20 failure times (in hours) collected during testing: 3691, 4123, 4808, 3542, 5654, 8231, 25883, 17124, 7784, 32027, 5219, 5674, 4146, 2794, 12230, 15401, 2999, 2893, 5121, 4557.

Because the gamma prior is conjugate to the exponential likelihood function, the posterior distribution will be a gamma distribution with parameters $\alpha + n$ and $\beta + \Sigma t_i$. In this example, $n = 20$ and $\Sigma t_i = 173{,}901$ h. We will use the OpenBUGS script in Table 11.19 to update the prior distribution for λ and calculate the predicted reliability over a 10,000 h mission time. We have also included the Bayesian model checks discussed in Chap. 4.

Running 100,000 iterations, after discarding 1,000 iterations for burn-in, we find the posterior mean of λ is 2.6×10^{-5}/h. The mean predicted reliability for 10,000 h is 0.77, with a fifth percentile of 0.73.

We are now at Step 5 in the Bayesian inference process, checking for problems in the model predictions. The simplest check is to look for obvious conflict between the prior distribution and the observed data. In this example, the prior distribution is very tightly peaked around a value for λ of 2×10^{-5}/h. It is so tightly peaked (see Fig. 11.2) that values much different from 2×10^{-5}/h are extremely unlikely. In marked contrast, the MLE of λ based on the 20 observed battery failure times is almost an order of magnitude larger, at about 1.2×10^{-4}/h. Because the posterior distributions is proportional to the product of the prior distribution and the likelihood function, the posterior distribution will give little weight to values of λ that are highly unlikely in the prior, as is the case here. This would appear to be a clear case of prior-data conflict.

The Bayesian model checks discussed in Chap. 4 confirm this conclusion. First, we can plot predicted vs. observed failure times, as in Fig. 11.3. This figure shows the observed failure times clustered well below the mean predicted failure time,

Table 11.19 OpenBUGS script for exponential aleatory model for battery failure example with conjugate gamma prior

```
model {
for(i in 1:N)    {
          time[i] ~ dexp(lambda)
          time.ranked[i] <- ranked(time[], i) #Order observed times
          time.rep[i] ~ dexp(lambda) #Replicate time from posterior
predictive distribution
          time.rep.ranked[i] <- ranked(time.rep[], i)
          F.obs[i] <- cumulative(time[i], time.ranked[i]) #CDF for observed ranked times
F.rep[i] <- cumulative(time.rep[i], time.rep.ranked[i]) #CDF for replicated ranked times
          diff.obs[i] <- pow(F.obs[i] - (2*i-1)/(2*N), 2)
          diff.rep[i] <- pow(F.rep[i] - (2*i-1)/(2*N), 2)
          }
CVM.obs <- sum(diff.obs[]) #Observed CVM statistic
CVM.rep <- sum(diff.rep[]) #Replicated CVM statistic
p.value <- step(CVM.rep - CVM.obs) #Should be near 0.5
time.pred ~ dexp(lambda)
#Predicted reliability at 10,000 hours
R.pred <- 1 - cumulative(time.pred, 1.E+4))
lambda ~ dgamma(52.2, 2.6E+6)
}
data
list(time=c(3691, 4123, 4808, 3542, 5654, 8231, 25883, 17124, 7784, 32027, 5219, 5674, 4146,
2794, 12230, 15401, 2999, 2893, 5121, 4557), N=20)
```

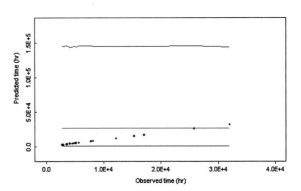

Fig. 11.3 Plot of observed failure times overlaid on 95% interval for predicted failure time suggests many observed times *near lower* bound of predicted failure time and most less than mean value. In other words, model appears to predict failure times that are generally longer than the observed ones

near the lower end of the 95% credible interval from the posterior predictive distribution.

Calculating the Bayesian p-value, we find a value of 0.0, confirming the model's inability to replicate the observed failure times with reasonable probability. The predictions of the model are clearly dominated by the prior.

At this point, the statistics suggest that we investigate the reason for the prior-data conflict. One thing to do might be to talk with the vendor to ascertain the basis

Fig. 11.4 Plot of observed failure times overlaid on 95% interval for predicted failure time for exponential model with Jeffreys prior suggests fewer observed times *near lower* bound of predicted failure time

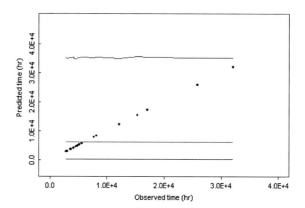

for the estimates provided for MTTF and the associated lower bound that were used to construct the prior distribution. Assume we learn that these estimates were based on tests of batteries in ideal, controlled environments. Also assume that it turns out that these test environments are different in salient ways from the environment anticipated for the 10,000 h battery mission.

We have convincing evidence that the vendor prior information is clearly inapplicable to the anticipated mission, although a simple exponential aleatory model may still be useful for reliability prediction. We can remove the influence of the dominating prior by replacing it with the Jeffreys noninformative prior, as discussed in Chap. 3. Recall that for exponential data, the Jeffreys prior is like a gamma distribution with both parameters equal to zero. We enter this in Open-BUGS as dgamma(0.0001, 0.0001), and provide an initial value of λ.

Running the script in Table 11.19 with the Jeffreys prior, we find a posterior mean for λ of 1.15×10^{-4}/h.[2] The mean predicted reliability for 10,000 h has decreased significantly to 0.33, and the fifth percentile value is 0.20. This significant drop in reliability illustrates the very strong influence the highly informative gamma prior above was having on the answer.

The plot of predicted times in Fig. 11.4 suggests no glaring problems with the exponential aleatory model for battery failure time. However, the Bayesian p-value for this model is 0.13, far enough from the ideal value of 0.5 to suggest looking for a more useful aleatory model, that is, one with better predictive ability.

Chapter 8 discussed several alternatives to the simple exponential model. One of these, which we will consider for this example, is a lognormal distribution. The OpenBUGS script in Table 11.20 implements a lognormal aleatory model, with diffuse priors on the parameters of the lognormal distribution.

Because this is a more complex aleatory model, with two parameters instead of one, we will run two chains in order to check for convergence, as described in Chap. 6. We find that the model converges quickly, within the first few hundred iterations. We will discard 1,000 iterations for burn-in to be safe. With this model

---

[2] With the Jeffreys prior, the posterior mean is numerically equal to the MLE.

Fig. 11.5 Plot of observed failure times overlaid on 95% interval for predicted failure time for lognormal aleatory model suggests no obvious problems with model

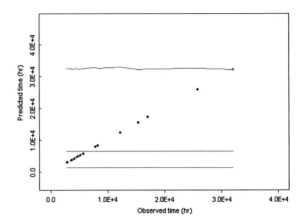

Table 11.20 OpenBUGS script for lognormal aleatory model for battery failure example

```
model {
for(i in 1:N) {
          time[i] ~ dlnorm(mu, tau)
          time.rep[i] ~ dlnorm(mu, tau)
          time.ranked[i] <- ranked(time[], i) #Order observed times
          time.rep.ranked[i] <- ranked(time.rep[], i)
          F.obs[i] <- cumulative(time[i], time.ranked[i]) #CDF for observed ranked times
          F.rep[i] <- cumulative(time.rep[i], time.rep.ranked[i]) #CDF for replicated ranked
          times
          diff.obs[i] <- pow(F.obs[i] - (2*i-1)/(2*N), 2)
          diff.rep[i] <- pow(F.rep[i] - (2*i-1)/(2*N), 2)
          }
CVM.obs <- sum(diff.obs[]) #Observed CVM statistic
CVM.rep <- sum(diff.rep[]) #Replicated CVM statistic
p.value <- step(CVM.rep - CVM.obs) #Should be near 0.5
time.pred ~ dlnorm(mu, tau)
#Predicted reliability at 10,000 hours
R.pred <- 1 - cumulative(time.pred, 1.E+4)
tau <- pow(sigma, -2)
sigma ~ dunif(0, 10)
mu ~ dflat()
} data
list(time=c(3691, 4123, 4808, 3542, 5654, 8231, 25883, 17124, 7784, 32027, 5219, 5674, 4146,
    2794, 12230, 15401, 2999, 2893, 5121, 4557), N=20)
inits
list(mu=8)
list(mu=9)
```

we find a mean predicted reliability for 10,000 h of 0.29, with a fifth percentile of 0.17. The plot of predicted failure times in Fig. 11.5 suggests no obvious problems, and this is confirmed by the Bayesian p-value of 0.29.

Fig. 11.6 Histogram of
observed failure times
suggests two clusters of times

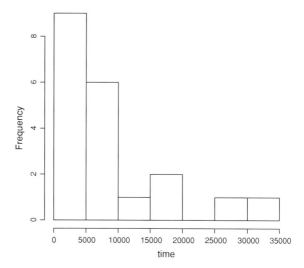

The Bayesian p-value for the lognormal model is better than the value of 0.13 for the simple exponential model, but is the additional complexity of the lognormal model (two parameters vs. one) justified? As discussed in Chap. 8, we can use the deviance information criterion (DIC), which OpenBUGS can calculate, to help decide this question. Recall that smaller values of DIC are indicative of better models. In this example, the DIC for the exponential aleatory model is 405, while that for the lognormal model is substantially lower, at 398. So we would conclude that the additional parameters in the lognormal model allow better predictive ability, without overfitting the observed data.

We could stop here, and use the reliability predictions of the lognormal model. However, let us ask whether this model could be improved further by including additional information. If we examine the observed failure times, we see there are a number of times on the order of a few thousand hours, some medium times of about 15,000 h, and some rather long times on the order of 30,000 h. The histogram in Fig. 11.6 supports the idea that there might be at least two clusters of failure times; with only 20 data points it is difficult to draw more conclusive insights.

Assume that in talking with the laboratories involved in producing the 20 test failure times, we learn that three different test environments were involved. We can use a regression model to enhance the lognormal aleatory model with this information.

The lognormal model is equivalent to a normal model for the logarithm of the failure times. The mean of the underlying normal model is the parameter we have designated as μ. We now let μ vary to capture the environmental influence on the test results: $\mu_i = \theta_1 + \theta_2 x_i$. In this equation, x_i measures the severity of the environment on a scale from 0 (least severe) to 1 (most severe). Assume that discussion with the test laboratories suggests three different environments and

Table 11.21 OpenBUGS script for lognormal aleatory model with environment included

```
model {
for(i in 1:N) {
          time[i] ~ dlnorm(mu[i], tau)
          time.rep[i] ~ dlnorm(mu[i], tau)
          mu[i] <- theta[1] + theta[2]*x[i]
          time.ranked[i] <- ranked(time[], i) #Order observed times
          time.rep.ranked[i] <- ranked(time.rep[], i)
          F.obs[i] <- cumulative(time[i], time.ranked[i]) #CDF for observed ranked times
          F.rep[i] <- cumulative(time.rep[i], time.rep.ranked[i]) #CDF for replicated ranked
          times
          diff.obs[i] <- pow(F.obs[i] - (2*i-1)/(2*N), 2)
          diff.rep[i] <- pow(F.rep[i] - (2*i-1)/(2*N), 2)
          }
CVM.obs <- sum(diff.obs[]) #Observed CVM statistic
CVM.rep <- sum(diff.rep[]) #Replicated CVM statistic
p.value <- step(CVM.rep - CVM.obs) #Should be near 0.5
x.pred <- 0.1
time.pred ~ dlnorm(mu.pred, tau)
mu.pred <- theta[1] + theta[2]*x.pred
#Predicted reliability at 10,000 hours
R.pred <- 1 - cumulative(time.pred, 1.E+4)
tau <- pow(sigma, -2)
sigma ~ dunif(0, 10)
theta[1] ~ dflat()
theta[2] ~ dflat()
}} data
list(time=c(3691, 4123, 4808, 3542, 5654, 8231, 25883, 17124, 7784, 32027, 5219, 5674, 4146,
     2794, 12230, 15401, 2999, 2893, 5121, 4557), N=20)
list(x=c(0.5, 0.5, 0.5, 0.9, 0.1, 0.5, 0.1, 0.9, 0.5, 0.1, 0.9, 0.9, 0.9, 0.9, 0.5, 0.1, 0.9, 0.9, 0.9, 0.1))
inits
list(theta=c(8, 0))
list(theta=c(9, -1))
```

associated x values of 0.1, 0.5, 0.9. Our new aleatory model for failure time is $T_i \sim$ dlnorm(μ_i, τ).

The OpenBUGS script is shown in Table 11.21. We have used independent diffuse priors on θ_1 and θ_2. Note also how the values of x for each failure time (supplied by the testers) have been added to the data statement.

To predict mission reliability for 10,000 h, we must make a judgment about the predicted environment, in terms of x. We have used a variable called x.pred in the script. If our value of x.pred is 0.5, the mean predicted reliability for 10,000 h is 0.31, with a 5th percentile of 0.18. The Bayesian p-value for this model is 0.71, about as far above 0.5 as the simple lognormal model was below it. The DIC for this model is 393, significantly better than the value of 398 for the simple lognormal model without the environment influence included.

The noteworthy point about the enhanced model is that it allows the environment to be factored into reliability predictions. For example, if the environment is expected to be quite severe, corresponding to x.pred = 0.9, say, we find the mean predicted reliability for 10,000 h to be only 0.12. On the other hand, for a quite benign environment (x.pred = 0.1), the mean predicted reliability is substantially higher, 0.59.

11.7 Summary

We have illustrated the value of developing models for unobservable parameters, such as O-ring failure probability, in which the unobservable parameter is a function of measurable parameters such as temperature and leak-test pressure. Incorporating such explanatory variables into the model helps to foster communication between risk analysts and system engineers, who are often more comfortable working with measurable quantities.

Bayesian estimation of the parameters in such models has been extremely difficult in the past, and has necessitated complex approximation methods, such as bootstrapping, to propagate parameter uncertainty through the model, the kinds of methods employed by [3] in their analysis. However, the advent of easy-to-use, powerful, open-source software such as OpenBUGS has made this type of analysis quite tractable, even for nonspecialists.

The Bayesian framework is particularly suited to risk-informed decision-making as it allows uncertainties in observable parameters such as temperature and other information to be incorporated into the model. The decision-maker can easily see the refinement in model estimates obtained by gathering additional information. The Bayesian framework is also well suited to model-checking, an important but often overlooked aspect of risk analysis. Model-checking also aids dialog between risk analysts and system engineers. In our O-ring example, system engineers thought leak-test pressure would be an important predictor of primary O-ring erosion, but this turned out not to be the case. This outcome could be fed back to the system engineers, and this could lead to additional dialog and model refinements. In the battery example, environment turned out to be an important variable, which can be incorporated into reliability predictions via a Bayesian regression model.

11.8 Exercises

1. The following random sample of 20 battery lifetimes has been collected: 3691, 4123, 4808, 3542, 5654, 8231, 25883, 17124, 7784, 32027, 5219, 5674, 4146, 2794, 12230, 15401, 2999, 2893, 5121, 4557.

(a) Using an exponential aleatory model for battery failure time, with the Jeffreys noninformative prior, find the posterior mean and 90% credible interval for the failure rate, λ.

(b) Using a replicate time from the posterior predictive distribution, find the posterior mean and 90% credible interval for the predicted battery reliability at 10,000 h.

(c) Consider an alternative aleatory model for the failure time that is lognormal(μ, σ^2), where μ is affected by battery quality in a linear fashion. In other words, $\mu = \theta_1 + \theta_2 Q$, where $0 < Q < 1$ is a measure of battery quality, with *larger* values of Q corresponding to *lower* quality. For the failure times above, the corresponding values of Q are 0.40, 0.36, 0.57, 0.97, 0.23, 0.51, 0.004, 0.77, 0.63, 0.06, 0.74, 0.80, 0.89, 0.74, 0.50, 0.08, 0.95, 0.92, 0.83, 0.20. Using independent, flat priors for θ_1 and θ_2, repeat part (b).

(d) Use posterior predictive plots to compare these two models.

(e) Use Bayesian p-value based on the Cramer-von Mises summary statistic to compare these two models.

(f) Use DIC to compare the relative fits of these two model

2. The Human Cognitive Reliability (HCR) model calculates operator non-response probabilities using a lognormal aleatory model for the response time T. The parameters of the model are the median response time, $t_{1/2}$, and the logarithmic standard deviation, σ. Consider the action of isolating a ruptured steam generator to limit the release of radioactivity. Table 3.3 in EPRI-TR-100259 estimates the mean of $t_{1/2}$ to be 500 s, with a range of 220–1021 s. Let us take this as prior information. A gamma distribution is a convenient functional form to represent the prior uncertainty in a positive variable such as $t_{1/2}$. We need two pieces of information to parametrize the gamma distribution, so we will use the mean and the upper end of the range, which we assume to be the 95th percentile. With this information, the gamma shape parameter is 3.3 and the scale parameter is 6.6×10^{-3}/s. For the prior on σ, we can use Table 3.1 in EPRI-TR-100259. We will take category CP2 for PWRs, for which the average is 0.38, with a 5/95 range of 0.07–0.69. The lognormal distribution gives a better representation of uncertainty over this wider range, so we will use a lognormal distribution with mean of 0.38 and error factor of 3.14. For data, we'll use the following isolation times. These will be assumed to be a random sample from a lognormal distribution. The times (in seconds) are 623, 719, 786, 799, 813, 817, 862, 909, 994, 1058, 1118, 1125, 1133, 1289.

(a) Find the posterior mean probability and 90% credible interval for the probability of not isolating the steam generator within 1200 s.

(b) Perform graphical and quantitative Bayesian model checks for this model.

(c) Develop a regression model that incorporates performance shaping factors that influence the median response time, $t_{1/2}$. Hint: what transformation of $t_{1/2}$ is appropriate?

References

1. Dezfuli H, Kelly DL, Smith C, Vedros K, Galyean W (2009) Bayesian inference for NASA probabilistic risk and reliability analysis. NASA, Washington, DC
2. McDonald AJ, Hansen JR (2009) Truth, lies, and O-rings: inside the space shuttle challenger disaster. University Press of Florida, FL
3. Dalal SR, Fowlkes EB, Hoadley B (1989) Risk analysis of the space shuttle: pre-challenger prediction of failure. J Am Stat Assoc 84(408):945–957
4. Hamada MS, Wilson AG, Reese CS, Martz HF (2008) Bayesian reliability. Springer, New York

Chapter 12
Bayesian Inference for Multilevel Fault Tree Models

12.1 Introduction

Fault tree modeling is an approach used to organize (in a graphical fashion) the various failure causes of a particular system. Typically, fault trees start with an undesired "top event," then combine the various ways the top event can occur using AND and OR operators (i.e., logic gates). A fault tree model is a static system representation that analyzes system failure behavior from a deductive logic point of view. That is, given a failure at a specific level in the fault tree, the question is asked: how could this failure have occurred?

A fault tree for a system is composed of "basic events" that are inputs to the logical gates. And, it is at this level of basic events where data and other information are commonly used, via Bayesian inference, to estimate component failure probabilities. In other words, the vast majority of fault tree analysis performed in support of PRA over the last 40 years has employed Bayesian inference at the *lowest level* in the fault tree model. However, information on system and component performance may be available at higher system "levels". For example, we might have failure information at the system, subsystem, or component levels. Also, we might have failure information for a group of components, where any one piece-part in this group can fail the group—this grouping is typically called a "super-component." We demonstrate how information at any level of a fault tree model can be combined via Bayesian inference to estimate failure probabilities at any level of that same fault tree model.

12.2 Example of a Super-Component with Two Piece-Parts

In this first example, assume we have a super-component that is placed in an operational demand situation, where we do not have any prior information on this component's behavior in such a setting. For illustration, we will assume the

D. Kelly and C. Smith, *Bayesian Inference for Probabilistic Risk Assessment*,
Springer Series in Reliability Engineering, DOI: 10.1007/978-1-84996-187-5_12,
© Springer-Verlag London Limited 2011

Table 12.1 OpenBUGS script for representing a super-component with only top-level failures represented

---

Single super-component, Jeffreys prior on p.TE
Demanded 10 times, No failures
model {
# Assign model for system top (TE) observable (x.TE number of failures)
x.TE ~ dbin(p.TE, n.TE)
Assume Jeffreys prior for TE failure probability
p.TE ~ dbeta(0.5, 0.5)
}
data
list(x.TE=0, n.TE=10)

Jeffreys prior on the component demand failure probability. We then collect operational data, where we demanded this super-component ten times and saw no failures.

As discussed in Chap. 3, the Jeffreys prior for a binomial aleatory failure model is a beta (0.5, 0.5) distribution, and the prior mean is $0.5/(0.5 + 0.5) = 0.5$. After performing the Bayesian update (using OpenBUGS) with this conjugate prior distribution, we find that the posterior failure probability mean value is 0.045. The script for this calculation is shown in Table 12.1.

This failure probability mean value (0.045) represents the "top level" of information (AKA, the super-component, or the "system"). Based upon the binomial aleatory model, and the associated data, we estimate the super-component performance is 0.045 (in this example, we focus on the mean value, but in reality we would have use of the entire posterior distribution for the failure probability).

Now, consider that we really had two piece-parts inside this single super-component. We would now like to estimate their respective failure probabilities. For the case of modeling failures at a lower level (where operational data is available), additional engineering information may be used to supplement the operational data. For example, assume a super-component is known to comprise multiple piece-parts. Further, we only have operational data for the super-component (e.g., we do not know which piece-part caused observed failures). It is possible to model the piece-parts *without* having direct observed operational data for these parts—and additional engineering information may be incorporated (part 1 is more reliable than part 2, part 3 is less reliable that part 1, etc.) into the modeling approach. Bayesian inference (with modern tools) is able to handle multiple sets of information simultaneously, regardless of the "level" of the modeling [1].

Returning to our two piece-part super-component, we can model the piece-parts as being a single system, where either part may fail the system (or super-component). The results of this simple system model (using OpenBUGS) would yield a posterior mean value (on a piece-part) of 0.024, where the applicable script is shown in Table 12.2.

Table 12.2 OpenBUGS script for a super-component which is made up of two piece-parts

```
model {
# Assume Jeffreys prior for subcomponent failure probability
p ~ dbeta(0.5, 0.5)
# Assign model for super-component
x.TE ~ dbin(p.TE, n.TE)
# Construct the overall fault tree structure (series system)
p.TE <- 1 - ((1-p)*(1- p))
}
data
list(x.TE=0, n.TE=10)
```

If we knew that each component was demanded on each of the ten super-component demands, then we would pool the data, so we have data consisting of no failures in 20 demands. This pooled data yields (using the Jeffreys prior) a mean failure probability (piece-part) of 0.024, the same as in the inference case above. This pooling of data for parts (at the lowest level of a fault tree) is the typical approach to PRA. However, we would need to perform the Bayesian calculation using the approach as described in the OpenBUGS script for this example if:

- We could not pool the data.
- We did not know the number of actual demands on the piece-parts.
- We wanted to incorporate other engineering information (e.g., part A is more reliable than part B).

12.3 Examples of a Super-Component with Multiple Piece-Parts

We could carry out the inference process for any number of piece-parts, for example if we had three piece-parts, we would see that the mean value (of a piece-part) was 0.016. Consequently, we can subdivide a super-component into n piece-parts in probability space and keep the same overall level of information at the top of the fault tree, that is, at the super-component level. It is important to note that in these OpenBUGS calculations, the entire distribution is estimated, not just the mean value.

In the example above, we encoded observational information at the super-component level, where in our example we did not see any failures at this level. In that situation, the information (no failures) is "shared" equally between the piece-parts since we did not include any additional engineering information (e.g., Part A is more reliable than Part B). If we see failures of specific piece-parts or if we have additional engineering information, these could be factored into the Bayesian inference process.

In the next example, assume we have a super-component with three piece-parts. However, we experienced a single failure in the ten demands. When a

Table 12.3 OpenBUGS script for a super-component with three piece-parts, where one has failed

---

3 piece-parts, Jeffreys prior on p_demand on each
Demanded 10 times, 1 failures on Component A
model {
# Assume Jeffreys prior for subcomponent failure probability
p ~ dbeta(0.5, 0.5)
p.B <- p # SOK dependence
p.C <- p # SOK dependence
x.B ~ dbin(p.B, n.B)
x.C ~ dbin(p.C, n.C)
p.A ~ dbeta(0.5, 0.5)
x.A ~ dbin(p.A, n.A)
Assign model for supercomponent
x.TE ~ dbin(p.TE, n.TE)
# Construct the overall fault tree structure
p.TE <- 1 - (1-p.A)*(1- p.B)*(1-p.C)
}
data
list(x.TE=1, n.TE=10, x.A=1, n.A=10, x.B=0, n.B=10, x.C=0, n.C=10)

---

super-component fails once in ten demands, its posterior mean failure probability is 0.14 (again, assuming we started with the Jeffreys prior).

Now, if we move to a lower level in the fault tree for the super-component and evaluate the three-piece-part case when Part A fails, we see the following posterior results (note that the prior at each level of the tree was assumed to be the Jeffreys prior):

- The failure probability mean value (piece-part A) = 0.11.
- The failure probability mean value (piece-part B, C) = 0.019.

The OpenBUGS script for this example is shown in Table 12.3.

Again, note that in this example, the overall super-component level information is preserved (its mean failure probability is 0.14) but Part A is much more unreliable than the other two parts, which is consonant with the information that has been collected.

In the case where we have additional engineering information, say at the piece-part level, this information can be factored into the Bayesian inference calculation. Assume we have the super-component (with three piece-parts) as in the last example. Again, we see a single failure (Part A) in ten demands of the super-component. However, for this example, assume that we had information that indicated (prior to collecting any data) that Part A was five times *more reliable* than B or C. For example, assume that we believed the prior failure probability for Part A was given by a beta(1, 9) distribution, which has a mean value of 0.1 (or five times lower than the mean of the prior for Part B and Part C, which are still using the Jeffreys prior).

Table 12.4 OpenBUGS script for a super-component with three piece-parts (one has failed), but where one part is believed to be more reliable than the others

```
model {
# Assume Jeffreys prior for the piece-part failure probability
p ~ dbeta(0.5, 0.5)
p.B <- p
p.C <- p # SOK correlation
x.B ~ dbin(p.B, n.B)
x.C ~ dbin(p.C, n.C)
p.A ~ dbeta(1, 9) # MORE reliable A component case (prior)
x.A ~ dbin(p.A, n.A)
# Assign model for supercomponent
x.TE ~ dbin(p.TE, n.TE)
# Construct the overall fault tree structure
p.TE <- 1 - (1-p.A)*(1- p.B)*(1-p.C)
}
data
list(x.TE=1, n.TE=10, x.A=1, n.A=10, x.B=0, n.B=10, x.C=0, n.C=10)
```

Using this informed prior for Part A, the Jeffreys prior for Parts B and C, and the one failure of Part A out of ten demands, we see the following posterior results:

- The mean value (piece-part A) = 0.091.
- The mean value (piece-part B, C) = 0.019.

The OpenBUGS script for this example is shown in Table 12.4.

Note that the posterior information for Part B and Part C has remained unchanged, but the mean failure probability for Part A is slightly lower, owing to its lower prior probability.

12.4 The "Bayesian Anomaly"

In the case where we are modeling various levels of a system, it is important to note that simplistic or naïve analysis can lead to misleading results. For example, one specific situation that has been referred to as the "Bayesian anomaly" focuses on the improper application of the multilevel modeling approach [2, 3].

To explore this issue, assume we have the super-component with three equal piece-parts. However, assume that we evaluated each of the three piece-parts independently from one another. Thus, an individual piece-part has a posterior failure probability mean (after ten demands, no failures) of 0.045.

Now, we wish to evaluate the next higher level, that of the super-component. So, we "OR" the three piece-parts (i.e., the super-component is failed if Part A fails or Part B fails or Part C fails). The super-component posterior mean then is:

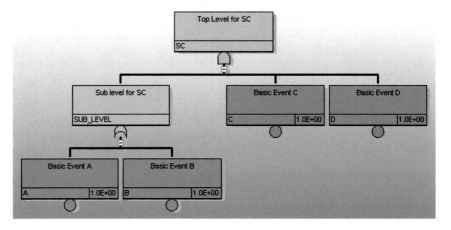

Fig. 12.1 Example super-component fault tree consisting of four sub-components

$$p(\text{super-component fails} \mid \text{no observed failures}) = 1 - (1 - 0.045)^3 = 0.13.$$

However, we know this to be incorrect (from our earlier calculation) since the super-component failure probability should be 0.045 (from Sect. 12.2). In effect, by treating each component as *knowledge-independent* of the other components, we are including additional information (e.g., from the Jeffreys priors for the three parts) that is not valid since the three parts share a state-of-knowledge correlation that is ignored via the simplistic analysis. As the number of piece-parts is increased, the simplistic analysis becomes even less accurate. Consequently, we need to be careful how high-level information is partitioned down to a lower level. However, moving from low-level information up to a higher level is not as problematic.

12.5 A Super-Component with Piece-Parts and a Sub-System

While we have demonstrated the ability to "flow down" data and information from a high-level (e.g., a super-component) to lower levels (e.g., piece-parts), the ability to represent multiple different levels in a complex system is a useful feature of Bayesian inference. To further illustrate this concept, let us explore a slightly more complicated example by assuming we have the super-component that is represented by the fault tree in Fig. 12.1. For this fault tree, the observed operational data are at multiple "levels," including the sub-system and the super-component level rather than just at the component level.

Further, assume we have data for the gate named "SUB_LEVEL" (the sub-system) and the top event labeled "SC." Specifically, assume there have been:

Fig. 12.2 Posterior distribution summaries for components B, C, and D (mean and 90% interval)

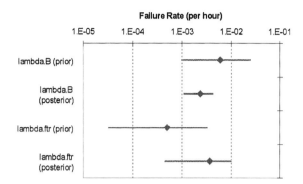

- Three failures of gate SUB_LEVEL in 20 demands (of sub-system level, not the entire super-component).
- One failure of the super-component in 13 demands of the system.

Note that the sub-system level (SUB_LEVEL) was also challenged during the 13 super-component demands (since the only way to have a super-component failure includes a component failure from this sub-subsystem). For the component level, assume:

- Basic events "A" and "B" represent a standby component, which must change state upon a demand.
- Assume that "A" represents failure to start and "B" represents failure to run for the required time (100 h).
- Basic events "C" and "D" represent normally operating components (100 h run time).

We will assume, for this calculation, the following prior lognormal distributions for the basic event parameters:

- Basic Event A: Mean 0.001 and Error Factor 5.
- Basic Event B: Median 0.005/h and Error Factor 5.
- Basic Event C: Median 0.0005/h and Error Factor 10.
- Basic Event D: Median 0.0005/h and Error Factor 10.

Note that since we use the same prior distribution to represent epistemic uncertainty for events C and D, we will have to account for this state-of-knowledge dependence in the Bayesian inference. The OpenBUGS script to perform this calculation is shown in Table 12.5.

Bayesian inference for this example yields a marginal posterior distribution for each of the four basic events. These posterior distribution sumaries show that the observed operational data has significantly increased the mean failure rate (lambda.ftr) for components C and D. The posterior mean of the failure rate (lambda.B) of component B has decreased from its prior mean value, while the mean failure probability for component A saw only a small change. An illustration of the components that did change is shown in Fig. 12.2.

Table 12.5 OpenBUGS script for the fault tree shown in Fig. 12.1

```
model    {
# This is system (fault tree top) observable (x.TE number of failures)
x.TE ~ dbin(p.TE, n.TE)
x.Gate.E ~ dbin(p.Gate.E, n.Gate.E)
p.TE <- p.Gate.E *p.C*p.D      # Probability of Top Event (from fault tree)
p.Gate.E <- p.A + p.B - p.A*p.B      # Probability of Gate E from fault tree
p.C <- p.ftr    # Account for state-of-knowledge dependence between C & D
p.D <- p.ftr    # by setting both to the same event
p.B <- 1 - exp(-lambda.B*time.miss)
p.ftr <- 1 - exp(-lambda.ftr*time.miss)
# Priors on basic event parameters
p.A ~ dlnorm(mu.A, tau.A)
mu.A <- log(mean.A) - pow(log(EF.A)/1.645, 2)/2
tau.A <- pow(log(EF.A)/1.645, -2)
lambda.B ~ dlnorm(mu.B, tau.B)
mu.B <- log(median.B)
tau.B <- pow(log(EF.B)/1.645, -2)
lambda.ftr ~ dlnorm(mu.ftr, tau.ftr)
mu.ftr <- log(mean.ftr) - pow(log(EF.ftr)/1.645, 2)/2
tau.ftr <- pow(log(EF.ftr)/1.645, -2)
}
data
list(x.TE=1, n.TE=13, x.Gate.E=3, n.Gate.E=20, time.miss=100)
list(mean.A=0.001, EF.A=5, median.B=0.005, EF.B=5, mean.ftr=0.0005, EF.ftr=10)
```

As illustrated in the previous example, we are able to mix both failure mechanisms (e.g., fails to start, fails to run) in addition to different levels of information in our Bayesian inference approach. For more details on this "multi-level" approach to Bayesian modeling see [4, 5].

12.6 Emergency Diesel Generator Example

To further demonstrate the general Bayesian approach, we will address another example related to modeling of a single emergency diesel generator (EDG). In the simplest case, one might represent the failure probability of an overall EDG as a constant for each test (even though the EDG is composed of hundreds of piece-parts). However, this assumption is not really accurate since different tests for the EDG may test different parts of the EDG in different ways. For example, some tests may only require a short run time, while others require a long run time. Further, the long run time will require additional demands on some components [e.g., fuel transfer pump (TP)] that may not be used during short tests. A simplified fault tree for the EDG in this example is shown in Fig. 12.3.

We have included three EDG failure modes in this fault tree:

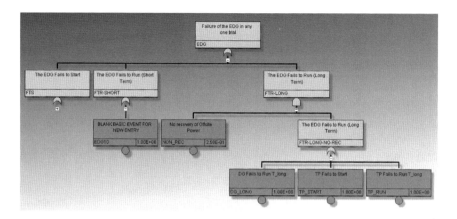

Fig. 12.3 Fault tree structure representing the EDG super-component and constituent components

1. Fails to start (FTS).
2. Fails to load/run (over a short duration) (FTR.S).
3. Fails to run (long duration) (FTR.L).

Assume that the time periods of interest (for the running failures are:

- t_short = 1 h.
- t_long = 24−1 = 23 h.

Assume that the EDG fuel oil transfer pump (TP) is demanded (exactly) six times during long tests. Note we could relax this assumption and model the number of demands probabilistically, as discussed in Chap. 10. Also, for simplicity we do not include offsite power recovery or repair from a failed state.

The uncertain parameter in the aleatory model for each basic event has an assumed prior distribution:

- lambda.short ~ gamma(0.495,164 h) (the CNI prior for failure to run).
- lambda.long ~ gamma(0.5, 625 h) (the CNI prior for failure to run).
- lambda.TP ~ gamma(0.5, 100,000 h) (the CNI prior for failure to run).
- p.TP ~ beta(0.498, 498) (the CNI prior for failure to start).

Then, in order to obtain the posterior distributions, we require operational data. Assume we have seen:

- 33 short tests (1 h duration) with no failures.
- 3 long tests (24 h) with 1 failure—TP failed to start.

Running this case with OpenBUGS (with the single TP failure) gives (posterior mean values):

- Lambda.TP = 5.0E-6/h (TP fails to run).
- Lambda.short = 2.5E-3/h (DG fails to run short term).

Fig. 12.4 Posterior density function for the EDG fault tree top event

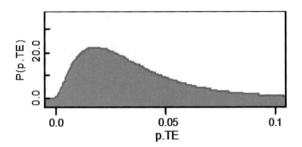

- Lambda.long = 7.5E-4/h (DG fails to run long term).
- p.TP = 2.9E-3/demand.
- EDG = 3.6E-2/test (p.TE: top event EDG fails), the marginal posterior density function for p.TE is shown in Fig. 12.4.

The top contributors to the failure probability of the EDG top event are (in order)

- p.TP_START = 1.7E-2.
- p.DG_LONG = 1.7E-2.
- p.DG_SHORT = 2.5E-3.
- p.TP_RUN = 1.1E-4.

12.7 Meeting Reliability Goals at Multiple Levels in a Fault Tree

We could extend any of these examples to determine to what degree we are meeting a reliability goal, set at any level of the fault tree. In order to demonstrate this concept, we will revisit the example from Sect. 12.3. In that example, we had a super-component with three piece-parts and we observed a single failure in ten demands. When the super-component fails once in ten demands, its failure probability posterior mean value is 0.14 (again, assuming we started with the Jeffreys prior). For the lower level in the fault tree, we found:

- The failure probability posterior mean value (piece-part A) = 0.11.
- The failure probability posterior mean value (piece-part B, C) = 0.019.

Assume that we had the following reliability goals:

- The system reliability is greater than 0.9.
- Each piece-part reliability is greater than 0.99.

The OpenBUGS script for this example is shown in Table 12.6:
The posterior results (monitoring the p.meet nodes) indicate:

- The probability of meeting our system reliability goal is about 33%.

Table 12.6 OpenBUGS script to determine whether reliability goals are met at different levels in a fault tree model

```
model {
# Assume Jeffreys prior for subcomponent failure probability
p ~ dbeta(0.5, 0.5)
p.B < - p # SOK dependence
p.C < - p # SOK dependence
x.B ~ dbin(p.B, n.B)
x.C ~ dbin(p.C, n.C)
p.A ~ dbeta(0.5, 0.5)
x.A ~ dbin(p.A, n.A)
# Assign model for supercomponent
x.TE ~ dbin(p.TE, n.TE)
# Construct the overall fault tree structure
p.TE <- 1 - (1-p.A)*(1- p.B)*(1-p.C)
# Compute goal nodes
TE.unrel.goal <- 1 - TE.goal
PP.unrel.goal <- 1 - PP.goal
p.meet.TE <- step(TE.unrel.goal - p.TE)
p.meet.PP.A <- step(PP.unrel.goal - p.A)
p.meet.PP.B <- step(PP.unrel.goal - p.B)
p.meet.PP.C <- step(PP.unrel.goal - p.C)
}
data
list(x.TE=1, n.TE=10, x.A=1, n.A=10, x.B=0, n.B=10, x.C=0, n.C=10)
list(TE.goal=0.9, PP.goal=0.99)
```

Fig. 12.5 Posterior density function for the fault tree top event showing the chance of meeting a reliability goal (shown in *green*)

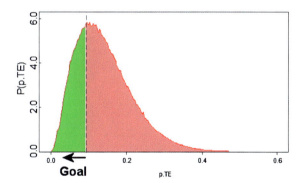

- The probability of meeting our piece-part reliability goal for A is about 2%.
- The probability of meeting our piece-part reliability goal for B is about 50%.
- The probability of meeting our piece-part reliability goal for C is about 50%.

The posterior distribution for the top event is shown graphically in Fig. 12.5, where the goal is placed at an unreliability of 0.1 (i.e., a reliability goal of 0.9

implies an unreliability of 0.1). The green area of the curve is the probability of meeting the reliability goal, which in this example is about 33%.

In this section, we described how information and data may be available at various levels in a fault tree model, and how these may be used within a Bayesian analysis framework to perform probabilistic inference on the model. Initially, we describe this approach using a simple fault tree model containing a single top event and two sub-events. The simple model represented a super-component and two piece-parts. For this system, we showed how modern Bayesian analysis tools (specifically the OpenBUGS software) may be used in a relatively straightforward manner to fully describe the probabilistic information known about the system. Then, we extend this approach to more complicated systems and sets of information.

Lastly, we showed how this same analysis tool (OpenBUGS) can also be used for the example models to determine the probability of meeting a reliability goal for any level in the fault tree model. In general, reliability goals are desirable system performance levels for driving improvements—not meeting specified goals is generally a flag indicating degraded performance.

References

1. Wilson A, Huzurbazar A (2006) Bayesian networks for multilevel system reliability. Reliab Eng Syst Saf 92:1413–1420
2. Guarro S, Yau M (2009) On the nature and practical handling of the Bayesian aggregation anomaly. Reliab Eng Syst Saf 94:1050–1056
3. Philipson L (2008) The 'Bayesian Anomaly' and its practical mitigation. IEEE Trans Reliab 57(1):171–173 March
4. Johnson V, Mossman A, Cotter P (2005) A hierachical model for estimating the early reliability of complex systems. IEEE Trans Reliab 54(2):224–231 June
5. Reese C, JohnsonV, Hamada M, Wilson A (2005) A hierachical model for the reliability of an Anti-Aircraft missile system. UT MD Anderson cancer center department of biostatistics working paper series, paper 9, October

Chapter 13
Additional Topics

This chapter begins by introducing Bayesian inference for extreme value processes, such as might be used to model high winds and flooding. It then gives an overview of the Bayesian treatment of expert opinion, and then proceeds to an example pointing out the pitfalls that can be encountered if *ad hoc* methods are employed. We next illustrate how to encode prior distributions into OpenBUGS that are not included as predefined distribution choices. We close this chapter with an example of Bayesian inference for a time-dependent Markov model of pipe rupture.

13.1 Extreme Value Processes

This section gives a brief introduction to statistical models for extreme quantities, such as flooding levels, which are encountered in external events PRA. Much of this material is adapted from [1], which gives an excellent overview of the subject from a mathematical level that should be comfortable to those with a degree in physical science or engineering, although with the exception of a quick overview of Bayesian inference in the last chapter, the treatment of inference in [1] is from a frequentist perspective.

As an example of the type of problem dealt with in this section, consider the following data on annual maximum sea levels, taken from [1]. From these data, one might wish to be able to project, with uncertainty, the maximum level for the next 100 or 500 years. A characteristic feature of this type of problem is the desire to extrapolate beyond the range of observed data.

The extrapolation is based on the so-called *extreme value paradigm*. It works by using asymptotic statistical models to make predictions. For the example above, if $X_1, X_2, \ldots$ are the sequence of daily maximum river levels, then we are interested in the maximum level for an observation period of n days:

D. Kelly and C. Smith, *Bayesian Inference for Probabilistic Risk Assessment*, Springer Series in Reliability Engineering, DOI: 10.1007/978-1-84996-187-5_13, © Springer-Verlag London Limited 2011

Fig. 13.1 Plot of annual sea
level in meters, from [1]

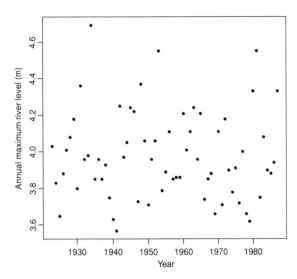

$$M_n = \max\{X_1, \ldots, X_n\} \qquad (13.1)$$

If the exact distribution of each daily maximum observation (X_i) were known,
then we could find the distribution of M_n. It is given by the following, under the
(perhaps dubious) assumption that the X_is are mutually statistically independent:

$$\Pr(M_n \le z) = \Pr(X_1 \le z, X_2 \le z, \ldots, X_n \le z)$$
$$= [F(z)]^n$$

In practice, we do not know the distribution of each X_i. We could estimate this
distribution, and find the distribution of M_n using the equation above, but small
errors in the estimate of F lead to large errors in F^n. However, we can find the
approximate distribution of M_n for large values of n, under certain assumptions.
We let $n \to \infty$, and this leads to a family of *extreme value distributions*, whose
parameters can be estimated from the observed data.

One can object to this procedure on the basis that the extrapolation to unseen
levels is faith-based; such a criticism is easy to make, and there is no real defense
against it, except to say that extrapolation is required, and using a method with some
rationale (asymptotic theory) is better than any existing alternatives. One must of
course be careful about gross violations of the underlying assumptions. For example,
what follows is based on the assumption that climate changes do not cause a systematic increase or decrease in the maximum annual river levels. Based on the plot in
Fig. 13.1, there does not appear to be any evidence of a change in the pattern of
variation over the observation period, but this is no guarantee for the future.

If you look at older references such as [2], you will find a somewhat confusing
set of limiting distributions, with names like the Gumbel distribution. Modern
references have unified all of these into a single family of limiting distributions,
referred to as the *generalized extreme value* (GEV) family of distributions.

For technical reasons, we cannot just look at the limiting behavior of M_n as $n \to \infty$. Instead, we look at $(M_n - a_n)/b_n$, where $\{a_n > 0\}$ and $\{b_n\}$ are sequences of constants that stabilize the location and scale of M_n as n increases, keeping the limiting distribution from becoming degenerate.

The big theorem is that $\Pr[(M_n - a_n)/b_n \le z] \to G(z)$ as $n \to \infty$, where $G(z)$ has the following form:

$$G(z) = \exp\left\{-\left[1 + \xi\left(\frac{z - \mu}{\sigma}\right)\right]^{-1/\xi}\right\} \tag{13.2}$$

$G(z)$ is defined for $1 + \xi(z - \mu)/\sigma > 0$. The parameters of the GEV distribution satisfy $-\infty < \mu < \infty$, $\sigma > 0$, and $-\infty < \xi < \infty$. The parameter ξ is a shape parameter and determines the tail behavior of $G(z)$.

Quantiles of $G(z)$ are often used, and are given by, for $G(z_p) = 1 - p$

$$z_p = \begin{cases} \mu - \dfrac{\sigma}{\xi}\left\{1 - [-\log(1 - p)]^{-\xi}\right\}, & \xi \neq 0 \\ \mu - \sigma\log[-\log(1 - p)], & \xi = 0 \end{cases} \tag{13.3}$$

The quantity z_p is often called the *return level* associated with the *return period* $1/p$; z_p is exceeded, on average, once every $1/p$ years (if we're measuring time in years). Put another way, z_p is exceeded by the annual maximum in any particular year with probability p.

If we define $y_p = -\log(1 - p)$, we can rewrite the quantiles as

$$z_p = \begin{cases} \mu - \dfrac{\sigma}{\xi}\left(1 - y_p^{-\xi}\right), & \xi \neq 0 \\ \mu - \sigma\log y_p, & \xi = 0 \end{cases} \tag{13.4}$$

If z_p is plotted against $\log(y_p)$, or z_p is plotted against y_p on a logarithmic scale, the plot will be linear if $\xi = 0$. If $\xi < 0$, the plot approaches the limit $\mu - \sigma/\xi$ as $p \to 0$. If $\xi > 0$, there is no upper bound to z_p. Such a graph is called a *return level plot*.

13.1.1 Bayesian Inference for the GEV Parameters

We treat the annual maximum river level data in Fig. 13.1 as a random sample from $G(z)$ ($n = 365$ is close to ∞). We will use OpenBUGS to perform the Bayesian inference for the GEV parameters. We will use independent diffuse priors: $\mu \sim$ dflat(), $\sigma \sim$ dgamma(10^{-4}, 10^{-4}), and $\xi \sim$ dflat(). The OpenBUGS script is shown in Table 13.1. The initial values for the three chains were centered on the maximum likelihood estimates of the parameters, which were obtained using the R package [3]. The posterior means and 95% credible intervals are shown in Table 13.2.

Table 13.1 OpenBUGS script for river level example

```
model {
for(i in 1:N) {
level[i] ~ dgev(mu, sigma, eta)
}
sigma ~ dgamma(0.0001, 0.0001)
mu ~ dflat()
xi ~ dflat()
}
data
list(level = c(4.03, 3.83, 3.65, 3.88, 4.01,
     4.08, 4.18, 3.8, 4.36, 3.96, 3.98, 4.69, 3.85, 3.96, 3.85, 3.93,
     3.75, 3.63, 3.57, 4.25, 3.97, 4.05, 4.24, 4.22, 3.73, 4.37, 4.06, 3.71, 3.96, 4.06, 4.55, 3.79, 3.89,
     4.11, 3.85, 3.86, 3.86, 4.21, 4.01, 4.11, 4.24, 3.96, 4.21, 3.74, 3.85, 3.88, 3.66, 4.11, 3.71, 4.18,
     3.9, 3.78, 3.91, 3.72, 4, 3.66, 3.62, 4.33, 4.55, 3.75, 4.08, 3.9, 3.88, 3.94, 4.33), N = 65)
inits
list(mu=3.8, sigma=1, xi=0)
list(mu=3.9, sigma=2, xi=-0.2)
list(mu=3.8, sigma=1.5, xi=0.1)
```

Table 13.2 Posterior summaries for GEV parameters in river level example

Table	Post mean	95% Interval
μ	3.87	(3.82, 3.93)
σ	0.20	(0.17, 0.25)
ξ	−0.03	(−0.21, 0.19)

A 100 year return level can be estimated as follows. This corresponds to $p = 0.01$ in Eq. 13.4. So we add the following line (based on Eq. 13.4) to the OpenBUGS script in Table 13.1:

$$\texttt{z.01 <- mu - sigma/xi*(1-pow (-log(1-0.01),-xi))}$$

The posterior mean is 4.8 m, with a 95% credible interval of (4.5, 5.4). A 500 year return level would be given by a similar line of script for `z.002`. The posterior mean for the 500 year return level is 5.1 m, with a 95% credible interval of (4.6, 6.2).

13.1.2 Thresholds and the Generalized Pareto Distribution

One objection to the use of annual maxima is that it is wasteful of data; we throw away all of the daily readings in each year except the largest one. Also, several values in one year may be larger than the maximum value in another year, but this information is lost as well. A way around these problems is to keep all of the data, and focus on the likelihood of exceeding some predetermined large threshold value.

If we treat X_1, X_2, ... as a sequence of independent and identically distributed variables, each with distribution F, then we can use the earlier GEV results as follows, and focus on the tail behavior of the GEV distribution. In other words, we want to find the conditional distribution of z, given that z is in the tail of the distribution, say $z > \mu$. First, we know that

$$\Pr(M_n \le z) = [F(z)]^n \approx G(z) = \exp\left\{-\left[1 + \xi\left(\frac{z-\mu}{\sigma}\right)\right]^{-1/\xi}\right\}$$

Taking logs of both sides gives

$$\log G(z) \approx -\left[1 + \xi\left(\frac{z-\mu}{\sigma}\right)\right]^{-1/\xi} \tag{13.5}$$

The Taylor series of $G(z)$ about μ is given by

$$G(z) \approx G(\mu) - g(\mu)(z - \mu) \approx 1 - g(\mu)(z - \mu)$$

where we have used $G(\mu) \approx 1$ because μ is assumed to be large.

The Taylor series for $\log[G(z)]$ is given by

$$\log[G(z)] \approx \log[G(\mu)] - \frac{g(\mu)}{G(\mu)}(z - \mu) \approx -g(\mu)(z - \mu)$$

And so we can write

$$\log[G(z)] \approx -[1 - G(z)]$$

Substituting into Eq. 13.5, we find

$$G(z) = 1 - \left[1 + \frac{\xi}{\sigma}(z - \mu)\right]^{-1/\xi} \tag{13.6}$$

The distribution in Eq. 13.6 is called a *generalized Pareto distribution* (GPD). The parameters (μ, σ, ξ) are the parameters of the GEV distribution discussed earlier. Our argument for deriving the GPD has been approximate, but a more rigorous argument leads to the same result. The tail of the GPD is bounded for $\xi < 0$, and unbounded for $\xi \ge 0$.

The data now consist of a sequence of independent and identically distributed measurements, z_1, z_2,..., z_n. We have to choose a threshold value (μ), and we then keep all z_is that are above this threshold value. The values above this threshold are treated as a random sample from a GPD with parameters μ, σ, and ξ.

As an example, the plot in Fig. 13.2 shows daily rainfall levels (in mm) for a period of 17,531 days, taken from [1]. Assume we have decided that a rainfall of 30 mm or more in a single day is of concern, so we set this as our threshold. The plot in Fig. 13.3 shows the rainfall values greater than this threshold.

Fig. 13.2 Plot of daily
rainfall in millimeter, from
[1]

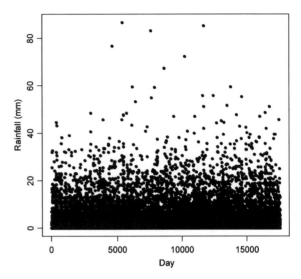

Fig. 13.3 Plot of daily
rainfall values from Fig. 13.2
in excess of 30 mm

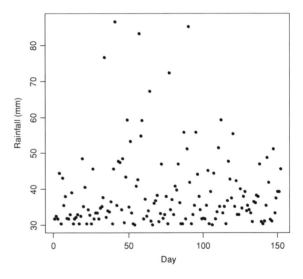

We model these excess values as a random sample from a GPD with parameters
$\mu = 30$, σ, and ξ. We use independent, diffuse priors as before: $\sigma \sim$ dgamma
$(10^{-4}, 10^{-4})$, and $\xi \sim$ dflat(). We could parse the data into values in excess of
30 mm in OpenBUGS, via a line of script such as rain.large[i] <- rain
[i]*step(rain[i] - 30). However, because the data file is so large, BUGS
runs *very* slowly if we do this. So instead, we created another data file with just the
rainfall values in excess of 30 mm. The OpenBUGS script is shown in Table 13.3.
The initial values for the two chains were centered on the maximum likelihood
estimates of the parameters, which were obtained using the R package [3]. The
posterior means and 95% credible intervals are shown in Table 13.4.

Table 13.3 OpenBUGS script for modeling excess rainfall data as GPD

```
model {
for(i in 1:N) {
rain.large[i] ~ dgpar(mu, sigma, xi)
}
mu <- 30
sigma ~ dgamma(0.0001, 0.0001)
xi ~ dflat()
}

list(sigma=7, xi=0.1)
list(sigma=7.5, xi=-0.1)
```

Table 13.4 Posterior summaries for GPD parameters in excess rainfall example

Parameter	Post. mean	95% Interval
σ	7.4	(5.7, 9.4)
ξ	0.21	(0.03, 0.44)

We can also calculate return levels with the GPD model. We are interested in finding z_m, the value that is exceeded on average every m days. Thus, for a 100 year return rainfall level, we would have (ignoring leap years) $m = (365)(100)$. We can write

$$\Pr(Z > z_m) = \Pr(Z > \mu)[1 - G(z_m)]$$

The first term in this equation we can denote as p_μ. The number of daily rainfall values $> \mu$ has a binomial distribution with parameters p_μ and n = number of daily rainfall values, which is 17,531 in our example. We will use the Jeffreys prior for p_μ, which is a beta $(0.5, 0.5)$ distribution.

We then substitute in $G(z)$, which is the GPD, given by Eq. 13.6:

$$\Pr(Z > z_m) = p_\mu \left[1 + \xi \left(\frac{z_m - \mu}{\sigma}\right)\right]^{-1/\xi} = \frac{1}{m}$$

Solving for z_m gives

$$z_m = \mu + \frac{\sigma}{\xi}\left[\left(mp_\mu\right)^{\xi} - 1\right]$$

The OpenBUGS script to calculate the 100 year return level is shown in Table 13.5. The posterior mean and 95% credible interval for the 100 year return level are 119 mm and (82.8, 201.2).

13.2 Treatment of Expert Opinion

The focus of this section is on methods for using information obtained from experts. Whether one has information from one or several experts, one would usually need to develop a representative estimate for use in the analysis. When

Table 13.5 OpenBUGS script to estimate 100 year return level for rainfall, based on GPD model

```
model {
for(i in 1:N) {
rain.large[i] ~ dgpar(mu, sigma, xi)
}
#100 year return level
x.large ~ dbin(p.mu, 17531)
p.mu ~ dbeta(0.5, 0.5)
m <- 100*365
z.m <- mu + (sigma/xi)*(pow(m*p.mu, xi) - 1)
mu <- 30
sigma ~ dgamma(0.0001, 0.0001)
xi ~ dflat()
}

data
list(x.large=152)

list(sigma=7, xi=0.1)
list(sigma=7.5, xi=-0.1)
```

the opinions of several experts are elicited, methods are needed to form the "aggregated" or "consensus" opinion. The formulation is quite simple conceptually. Expert opinion is simply treated as information about the unknown parameter of interest. The information is then used to update the analyst's own (prior) estimate through Bayes theorem. We will describe some of the basic techniques for a number of important classes of problems, but the coverage will not be exhaustive as the techniques for certain classes of problems tend to become very complicated without any assurance of significant improvement in the resulting estimates.

13.2.1 Information from a Single Expert

In this case the expert provides an estimate for an unknown parameter of interest, such as λ in the Poisson distribution. To use this information to update the analyst's prior distribution for λ via Bayes' Theorem, a likelihood function is needed for the information obtained from the expert. When the epistemic uncertainty in parameter values spans several orders of magnitude, as is common in PRA, a lognormal distribution is a convenient likelihood function. The parameter τ (logarithmic precision) in the lognormal distribution will represent the analyst's assessment of the expert's expertise: small values of τ correspond to low confidence (high uncertainty) and vice versa. A bias factor can also be introduced if desired, with bias less than one meaning the analyst believes the expert tends to

Table 13.6 OpenBUGS script for lognormal (multiplicative error) model for information from single expert

```
model {
lambda.star ~ dlnorm(mu, tau) # Lognormal likelihood for information from expert
lambda.star <- 1/MTTF
mu <-log(lambda*bias)
tau <- pow(log(EF.expert)/1.645, -2)
lambda ~ dlnorm(mu.analyst, tau.analyst) # Analyst's lognormal prior for lambda
mu.analyst <- log(prior.median)
tau.analyst <- pow(log(prior.EF)/1.645, -2)
}
data
list(MTTF=500000, bias=0.75, EF.expert=10, prior.median=0.000001, prior.EF=10)
```

underestimate the true value of τ, and a factor greater than one means the expert tends to overestimate the true value.

As an example, assume the analyst's estimate for the failure of a level sensor is 10^{-6}/h, but the analyst is not very confident of this estimate. The analyst adopts a lognormal distribution with this estimate as the median, and an error factor of 10 to describe the uncertainty. The level sensor vendor (the "expert") provides an estimate of the mean time to failure (MTTF) for the level sensor. The vendor's estimate for the MTTF is 500,000 h. We can develop a posterior distribution for the level sensor failure rate that uses these two sources of information.

The first step is to convert the MTTF estimate provided by the vendor into an estimate of the failure rate. This can be done by taking the reciprocal of the MTTF estimate. The analyst must assess an uncertainty factor on the vendor's estimate, representing their confidence in the estimate provided by the vendor. Assume that the analyst is not very confident in the vendor's estimate, and assesses an error factor of 10. He also believes that the expert tends to overestimate the MTTF, that is, he *underestimates* the failure rate, so a bias factor of 0.75 is assessed by the analyst. The OpenBUGS script in Table 13.6 is used to analyze this problem. Running the script in the usual way gives a posterior mean for λ of 2.7×10^{-6}/h with a 90% credible interval of $(3.2 \times 10^{-7}, 8.4 \times 10^{-6})$. If the analyst thought the vendor tended to underestimate the MTTF, that is, overestimate λ, he would use a bias factor greater than one. Changing the bias factor to 5, for example, changes the posterior mean to 1.0×10^{-6}/h with a 90% credible interval of $(1.3 \times 10^{-7}, 3.2 \times 10^{-6})$.

13.2.2 Using Information from Multiple Experts

Cases encountered in practice often involve more than one expert. When multiple experts are involved the main question concerns the method of aggregation or pooling to form a single representative or aggregate estimate from the multiple expert estimates. A number of *ad hoc* approaches have been used for combining information from multiple experts, such as taking the geometric average (arithmetic average

Table 13.7 Expert estimates of pressure transmitter failure rate, from [4]	Expert	Estimate (per hour)	Confidence measure (error factor)
	1	3.0×10^{-6}	3
	2	2.5×10^{-5}	3
	3	1.0×10^{-5}	5
	4	6.8×10^{-6}	5
	5	2.0×10^{-6}	5
	6	8.8×10^{-7}	10

Table 13.8 OpenBUGS script for combining information from multiple experts using multiplicative error model (lognormal likelihood)

```
model    {
for(i in 1:N){
          lambda.star[i] ~ dlnorm(mu, tau[i])
          tau[i] <- pow(log(EF[i])/1.645, -2)
          }
mu ~ dflat() # Diffuse prior on mu
lambda <- exp(mu) # Monitor this node for aggregate distribution
}


data
list(lambda.star=c(3.E-6, 2.5E-5, 1.E-5, 6.8E-6, 2.E-6, 8.8E-7), EF=c(3,3,5,5,5,10), N=6)
inits
list(mu=-10)
list(mu=-5)
```

of the logarithms) and taking the low and high estimates as the 5 and 95th percentiles of a lognormal distribution. A justification commonly given for these *ad hoc* approaches is that analytical techniques need not be more sophisticated than the pool of estimates (experts' opinions) to which they are applied. Therefore, a simple averaging technique (equal weights) has often been judged satisfactory as well as efficient, especially when the quantity of information collected is large.

The Bayesian approach of the previous section can be expanded to include multiple experts. Basically, the hierarchical Bayes methods of Chap. 7 can be used to develop a prior distribution representing the variability among the experts. While mathematically cumbersome, this is straightforward to encode in Open-BUGS, as the following example from [4] illustrates.

Six estimates are available for the failure rate of pressure transmitters. These estimates along with the assigned measure of confidence are listed in Table 13.7. The analyst wishes to aggregate these estimates into a single distribution that captures the variability among the experts.

As in the previous section, the likelihood function for each expert will be assumed to be lognormal. A diffuse prior is placed on λ (actually on the logarithm of λ). The OpenBUGS script in Table 13.8 is used to develop a distribution for λ, accounting for the variability among the six experts. Running the script in the

Table 13.9 Hypothetical failure data for fan check valves, from [5]

Record number	Failure mode	Failures	Demands
469	FTO	0	11,112
470	FTO	0	3,493
471	FTO	0	10,273
472	FTO	1	971
473	FTO	0	4,230
474	FTO	0	704
475	FTO	0	7,855
476	FTO	0	504
477	FTO	0	891
478	FTO	0	846
480	FTO	0	572
481	FTO	0	631
482	FTO	0	2,245
488	FTO	0	7,665
532	FTO	0	1,425
534	FTO	0	700
538	FTO	0	716
549	FTO	8	1,236
550	FTO	0	926
551	FTO	1	588
552	FTO	0	856
554	FTO	1	708
569	FTO	0	724
570	FTO	12	8,716
592	FTO	2	632
593	FTO	0	564

usual way gives a posterior mean for λ of 6.5×10^{-6}/h, with a 90% credible interval of $(3.4 \times 10^{-6}, 1.1 \times 10^{-5})$.

13.3 Pitfalls of *ad hoc* Methods

For a group of failure records, one *ad hoc* technique that has been encountered by the authors is a type of data pooling that attempts to approximate the hierarchical Bayes approaches of Chap. 7. For example, each source might be used to generate a mean and variance as follows. If the number of failures (x) is greater than zero, then each source can be used to generate a beta distribution for p with parameters $\alpha = x$ and $\beta = n - x$, where n is the number of demands (the standard conjugate updating approach). From the properties of a beta distribution, the mean is then x/n and the variance is approximately $x(n - x)/n^3$. If no failures were recorded for a particular data source, then α might be taken to be 0.5 (assuming the Jeffreys prior for p is used). For the hypothetical data in Table 13.9, taken from [5], the overall mean and variance can be found to be

Table 13.10 Summary of overall fan check valve prior, average-moment approach

Fitted distribution	Mean	5th Percentile	95th Percentile
Beta	9.6E-4	9.4E-8	4.0E-3
Lognormal	9.6E-4	6.8E-5	3.3E-3

9.6E-4 and 2.8E-6, respectively. From these moments, the parameters of the resulting beta distribution can be found using:

$$\alpha_{tot} \approx \frac{\mu_{tot}^2}{\sigma_{tot}^2}$$

$$\beta_{tot} = \frac{\alpha_{tot}(1 - \mu_{tot})}{\mu_{tot}}$$

A lognormal distribution could also be fit using these overall moments. For the data in Table 13.9, the overall beta distribution has parameters 0.3 and 343.6. The fitted lognormal distribution has mean 9.6E-4 and error factor (EF) of 7.

The overall posterior developed by this *ad hoc* method using the average-moment approach depends on the fitted distribution and is summarized below for two different distributions Table 13.10.

Because of the large source-to-source variability exhibited by the data in Table 13.9, it may be inappropriate to pool the data as was done above. The standard Bayesian approach to such a problem is to specify a hierarchical prior for the demand failure probability, p, as described in Chap. 7. We will compare the results from this approach with the *ad hoc* average-moment approach above. We will analyze two different first-stage priors, beta and logistic-normal, with independent diffuse hyperpriors in both cases. The OpenBUGS script in Table 13.11 was used to carry out the analysis.

13.3.1 Using a Beta First-Stage Prior

The overall average distribution representing source-to-source variability in p has a mean of 1.2×10^{-3}, variance of 1.4×10^{-4}, and a 90% credible interval of $(3.5 \times 10^{-20}, 4.1 \times 10^{-3})$. The very small 5th percentile is an artifact of choosing a beta distribution as a first-stage prior. The posterior mean of α is 0.12, and the average variability distribution has a sharp vertical asymptote at $p = 0$.

13.3.2 Using a Logistic-Normal First-Stage Prior

The logistic-normal distribution is constrained to lie between 0 and 1, and because the density function goes to 0 at both 0 and 1, it avoids the vertical asymptote at $p = 0$ from which the above beta distribution suffers. For small values of p, the logistic-normal and lognormal distributions are very close; we chose to use the

Table 13.11 OpenBUGS hierarchical Bayes analysis of data in Table 13.9

```
model {
for(i in 1:N)  {
        x[i]  ~  dbin(p[i], n[i]) # Binomial model for number of events in each source
        p[i]  ~  dbeta(alpha, beta) # First-stage beta prior
        #p[i]  <- exp(p.norm[i])/(1 + exp(p.norm[i])) # Logistic-normal first-stage prior
        #p.norm[i]  ~  dnorm(mu, tau)
        x.rep[i]  ~  dbin(p[i], n[i]) # Replicate value from posterior predictive distribution
        #Generate inputs for Bayesian p-value calculation
        diff.obs[i]  <- pow(x[i] - n[i]*p[i], 2)/(n[i]*p[i]*(1-p[i]))
        diff.rep[i]  <- pow(x.rep[i] - n[i]*p[i], 2)/(n[i]*p[i]*(1-p[i]))
        }
p.avg  ~  dbeta(alpha, beta)     #Average beta population variability curve
#p.avg  ~  dlnorm(mu, tau)
#p.norm.avg  ~  dnorm(mu, tau)
#p.avg  <- exp(p.norm.avg)/(1 + exp(p.norm.avg))
#Compare observed failure total with replicated total
x.tot.obs  <- sum(x[])
x.tot.rep  <- sum(x.rep[])
percentile  <- step(x.tot.obs - x.tot.rep)  #Looking for values near 0.5
# Calculate Bayesian p-value
chisq.obs  <- sum(diff.obs[])
chisq.rep  <- sum(diff.rep[])
p.value  <- step(chisq.rep - chisq.obs)  #Mean of this node should be near 0.5
# Hyperpriors for beta first-stage prior
alpha  ~  dgamma(0.0001, 0.0001)
beta  ~  dgamma(0.0001, 0.0001)
#mu  ~  dflat()
#tau  <- pow(sigma, -2)
#sigma  ~  dunif(0, 20)
}

inits
list(alpha=1, beta=100) #Chain 1
list(alpha=0.1, beta=200) #Chain 2
list(mu=-11, sigma=1)
list(mu=-12, sigma=5)
```

logistic-normal distribution because, with such large variability, the Monte Carlo sampling in OpenBUGS can generate values of $p > 1$, and these have the potential to skew the results, particularly the mean.

With a logistic-normal first-stage prior, we found the overall average distribution representing source-to-source variability to have a mean of 0.01, variance of 8.7E-3, and 90% credible interval of (7.6E-11, 1.2E-2).

Table 13.12 Posterior results for the *ad hoc* versus Bayesian method comparison

Method	Mean	5th Percentile	95th Percentile
Ad hoc (beta)	1.3E-4	1.4E-8	5.9E-4
Ad hoc (lognormal)	3.0E-4	4.3E-5	8.3E-4
Hierarchical Bayes (beta)	4.9E-5	5.9E-23	2.9E-4
Hierarchical Bayes (logistic-normal)	4.7E-5	2.6E-11	2.5E-4
Jeffreys prior	2.5E-4	9.8E-7	9.6E-4

Fig. 13.4 Plot of the posterior results for the Bayesian versus *ad hoc* method comparison

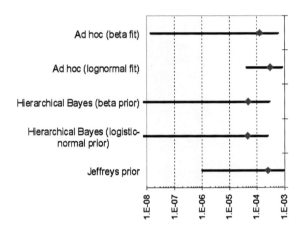

13.3.3 Update with New Data

Assume we would like to update the overall check valve prior with new data, which we take to be 0 failures in 2,000 demands. The results of the four different update possibilities are shown in Table 13.12. As a reference point, we include an update of the Jeffreys prior.

As shown in Fig. 13.4, a perhaps surprising outcome is that updating the lognormal distribution fit with the *ad hoc* average-moment approach gives about the same mean and 95th percentile for p as simply updating the Jeffreys prior (however, the 5th percentile differs between the two posteriors). The beta prior would give about the same result if there were not a vertical asymptote at $p = 0$, causing excess shrinkage of the mean toward 0.

Both hierarchical Bayes analyses give similar means and 95th percentiles; the 5th percentiles differ because of the vertical asymptote in the beta first-stage prior. Hierarchical Bayes allows the large number of sources with zero failures to more strongly influence the result than the average-moment *ad hoc* approach. With no failures in 2,000 demands, the posterior mean is pulled more towards a value of zero in the hierarchical Bayes analysis, giving a less conservative result.

Table 13.13 Model validation results for the *ad hoc* versus Bayesian method comparison

Method	Total replicated failures (mean)	Bayesian p-value
Ad hoc (beta)	61.0	0.001
Ad hoc (lognormal)	67.3	0.001
Hierarchical Bayes (beta)	25.1	0.44
Hierarchical Bayes (logistic-normal)	25.0	0.38

13.3.4 Model Checking

We can generate replicate failure counts for the data sources in Table 13.9, and then use the Bayesian p-value calculated from the chi-square summary statistic described in Chap. 4 to compare models. Table 13.13 shows the results of this model-checking calculation. The *ad hoc* distributions derived from the average-moment approach are poor at replicating the observed data: they over-predict the total number of failures (the observed total was 25) and they under-predict the variability in the failure count, leading to a very low Bayesian p-value. In contrast, the hierarchical Bayes models have much better predictive validity.

13.4 Specifying a New Prior Distribution in OpenBUGS

There are times when an analyst may wish to use a distribution that is not available directly as a choice in OpenBUGS. We have already encountered two instances of this. In Chap. 3, we say how to specify a logistic-normal prior for p in the binomial distribution by specifying the underlying normal distribution and then transforming. In Chap. 9 we saw how to specify an aleatory model for failure with repair in the case of a power-law nonhomogeneous Poisson process. There, we used the dloglik() distribution, which requires us to specify the logarithm of the likelihood function. In this section, we show how to use the dloglik() distribution to enter a prior distribution, where there is no underlying distribution to exploit via a transformation, as we were able to do in the case of the logistic-normal distribution.

Suppose that the analyst wishes to use a truncated exponential distribution for p in a binomial aleatory model, this being a type of maximum entropy prior for p, if the analyst knows a mean value, μ, and lower and upper bounds a and b, respectively. As discussed in [6], the density function is given by

$$f(p) = \frac{\beta e^{\beta p}}{e^{\beta b} - e^{\beta a}}$$

where β is determined by the specified mean constraint, μ. As an example, assume that p is known to lie between $a = 0.1$ and $b = 0.8$. Assume that the mean is specified as $\mu = 0.7$. The parameter β is found to be 4.5 by numerically solving the following equation:

Table 13.14 OpenBUGS script to specify maximum entropy prior for *p*

```
model {
p ~ dunif(0.1, 0.9)
zero <- 0
zero ~ dloglik(phi)
phi <- log(beta) + beta*p - log(exp(beta*b) - exp(beta*a))
beta <- solution(F(s), 2, 8, 0.1)
F(s) <- (b*exp(s*b) - a*exp(s*a))/(exp(s*b) - exp(s*a)) - 1/s - mu
x ~ dbin(p, n)
}
data
list(a=0.1, b=0.9, mu=0.7)
list(x=5, n=9)
```

$$\frac{be^{\beta b} - ae^{\beta a}}{e^{\beta b} - e^{\beta a}} - \frac{1}{\beta} - \mu = 0$$

The OpenBUGS script in Table 13.14, solves for β using the solution() function, specifies the maximum entropy prior for p, and updates it with 5 events in 9 trials, giving a posterior mean for p of 0.63, which we note lies between the prior mean of 0.7 and the MLE of 0.56.

13.5 Bayesian Inference for Parameters of a Markov Model

Markov models are occasionally encountered in PRA applications, especially when time-dependence is an explicit concern. In this section, we illustrate the capability to simultaneously perform Bayesian inference for the Markov model parameters and solve the system of Markov ordinary differential equations (ODEs) within OpenBUGS.

We take as our example the Markov model used in [7] to estimate piping rupture frequency. This model is shown in Fig. 13.5.

13.5.1 Aleatory Models for Failure

The occurrences of failures as a result of stress corrosion cracking (SC) and design and construction defects (DC) were assumed to be described by independent Poisson processes, with rates λ_{SC} and λ_{DC}, respectively. The occurrence of failures overall is then described by a Poisson process with rate $\lambda = \lambda_{SC} + \lambda_{DC}$. Data consist of the number of SC and DC failures, n_{SC} and n_{DC}, observed over specified exposure times.

Reference [7] accounted for uncertainty in the exposure times via a discrete distribution, with nine components for SC, and three for DC. Lognormal priors for

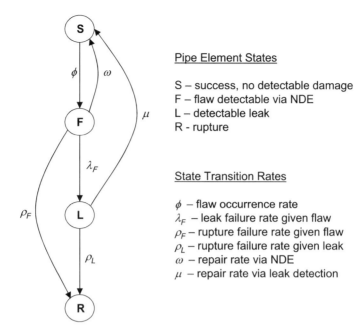

Fig. 13.5 Markov model for piping rupture, taken from [7]

λ_{SC} and λ_{DC} were updated with $n_{SC} = 8$ and $n_{DC} = 4$, with the exposure times and weights given in Fig. 13.6.

The posterior distributions for λ_{SC} and λ_{DC} are obtained by averaging over the weights given in Fig. 13.6. The overall failure rate, λ, is then found by summing λ_{SC} and λ_{DC}. The posterior distribution for λ is multimodal.

Occurrences of ruptures conditional upon failures were described by a binomial distribution with parameters $P(R_1)$ and 12 (sum of n_{SC} and n_{DC}). Each $P(R_1)$ has a lognormal prior distribution, as given in Table 13.15.

The frequency of pipe rupture of a given size is found by multiplying λ by $P(R_1)$.

13.5.2 Other Markov Model Parameters

Uncertainties for the other parameters were represented as described in [7], with the exception of P_{FD} and P_{LD}, for which [7] used a triangular distribution. Because the triangular distribution is not implemented in OpenBUGS, a beta distribution was used over the range given in [7], with a mean value approximately equal to the mode of the triangular distribution.

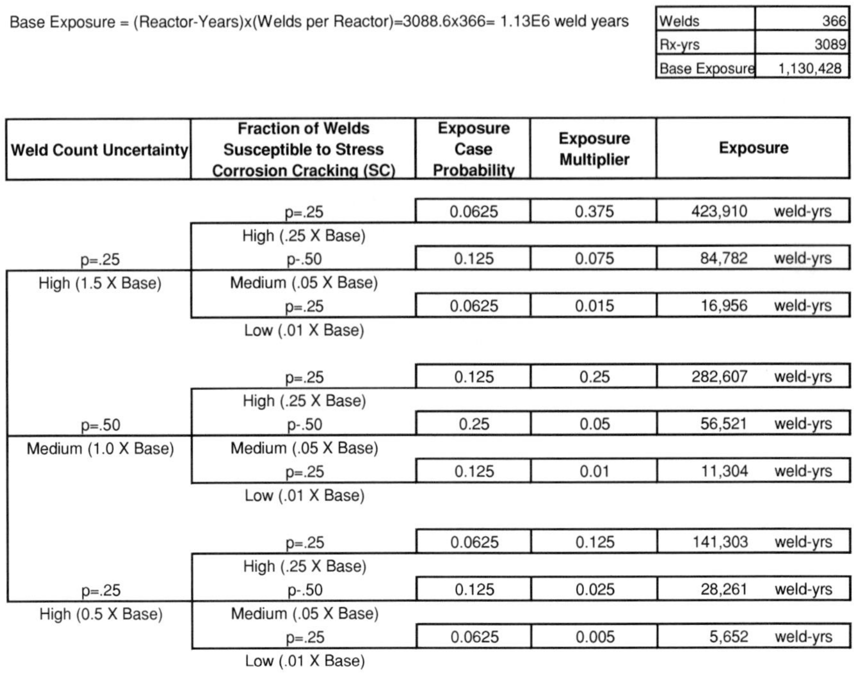

Base Exposure = (Reactor-Years)x(Welds per Reactor)=3088.6x366= 1.13E6 weld years

Welds	366
Rx-yrs	3089
Base Exposure	1,130,428

Weld Count Uncertainty	Fraction of Welds Susceptible to Stress Corrosion Cracking (SC)	Exposure Case Probability	Exposure Multiplier	Exposure	
p=.25 High (1.5 X Base)	p=.25 High (.25 X Base)	0.0625	0.375	423,910	weld-yrs
	p=.50 Medium (.05 X Base)	0.125	0.075	84,782	weld-yrs
	p=.25 Low (.01 X Base)	0.0625	0.015	16,956	weld-yrs
p=.50 Medium (1.0 X Base)	p=.25 High (.25 X Base)	0.125	0.25	282,607	weld-yrs
	p=.50 Medium (.05 X Base)	0.25	0.05	56,521	weld-yrs
	p=.25 Low (.01 X Base)	0.125	0.01	11,304	weld-yrs
p=.25 High (0.5 X Base)	p=.25 High (.25 X Base)	0.0625	0.125	141,303	weld-yrs
	p=.50 Medium (.05 X Base)	0.125	0.025	28,261	weld-yrs
	p=.25 Low (.01 X Base)	0.0625	0.005	5,652	weld-yrs

Fig. 13.6 Discrete distribution model for service data exposure uncertainty

Table 13.15 Lognormal prior distributions for pipe rupture probabilities conditional upon pipe failure

Symbol	Break size (in.)	Lognormal prior distribution Mean	Range factor
$P(R_1)$	0.032	0.13	2.8
$P(R_2)$	0.10	0.06	3.4
$P(R_3)$	0.32	0.021	4.5
$P(R_4)$	1.00	5.00E-03	6.6
$P(R_5)$	3.16	8.00E-04	10.5
$P(R_6)$	10.0	1.80E-04	15.1
$P(R_7)$	31.62	3.80E-05	21.8
$P(R_8)$	42.4	2.11E-05	25.1

13.5.3 Markov System Equations

The Markov model is described by a set of four coupled linear first-order ordinary differential equations (ODEs). The equations give the rate of change of the four components of the state probability vector in terms of the Markov parameters and the state probability vector components, which are time-dependent. The initial condition necessary for the solution of the vector equation is $P(0) = (1, 0, 0, 0)^{\mathrm{T}}$.

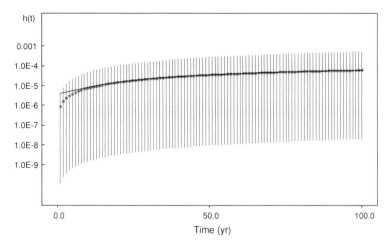

Fig. 13.7 System hazard rate as a function of time for 10 inch rupture. Line is posterior mean and bars illustrate 95% credible intervals

The system of ODEs can be written as

$$A = \begin{pmatrix} -\phi & \omega & \mu & 0 \\ \phi & -(\lambda_F + \rho_F + \omega) & 0 & 0 \\ 0 & \lambda_F & -(\rho_L + \mu) & 0 \\ 0 & \rho_F & \rho_L & 0 \end{pmatrix}$$

With rupture (state 4) defined as failure for the system, the system hazard rate is given by

$$h(t) = \frac{f(t)}{1 - F(t)} = -\frac{1}{r(t)}\frac{dr}{dt}$$
$$= \frac{\rho_F P_2(t) + \rho_L P_3(t)}{\sum_{i=1}^{3} P_i(t)}$$

13.5.4 Implementation in OpenBUGS

In OpenBUGS, a numerical ODE solver is used, via the ode() function. The OpenBUGS script in Table 13.16 implements this example. Note the use of a transformation to encode a beta distribution over a range other than [0, 1]. The model converged quickly, but for conservatism the first 1,000 iterations were discarded for burn-in. Parameter estimates are based on 50,000 iterations after burn-in, resulting in a Monte Carlo error of 1% or less of the overall standard deviation for each parameter. Posterior summaries for the input parameters

Table 13.16 OpenBUGS script for Markov pipe rupture model

```
model {
#Aleatory models for number of failures
for(i in 1:9) {
          n.sc[i] ~ dpois(mean.sc[i])
          mean.sc[i] <- lambda.sc[i]*time.sc[i]
          lambda.sc[i] ~ dlnorm(mu.sc, tau.sc)
          }
for(j in 1:3) {
          n.dc[j] ~ dpois(mean.dc[j])
          mean.dc[j] <- lambda.dc[j]*time.dc[j]
          lambda.dc[j] ~ dlnorm(mu.dc, tau.dc)
          }
for(k in 1:8) {
          x.R[k] ~ dbin(p.R[k], n.R)
          p.R[k] ~ dlnorm(mu.R[k], tau.R[k])
          mu.R[k] <- log(prior.mean.R[k]) - pow(sigma.R[k], 2)/2
          sigma.R[k] <- log(RF.R[k])/1.645
          tau.R[k] <- pow(sigma.R[k], -2)
#rho.rupt[k] is used to calculate the transition rate rho.F below
          rho.rupt[k] <- lambda*p.R[k]
          }
##################################
#Weighted-average posterior distributions
lambda.sc.avg <- lambda.sc[r.sc]
r.sc ~ dcat(w.sc[])
lambda.dc.avg <- lambda.dc[r.dc]
r.dc ~ dcat(w.dc[])
lambda <- lambda.sc.avg + lambda.dc.avg
######################################
#Other Markov model parameters
phi <- m.f*lambda
m.f.trunc ~ dbeta(1, 2)
m.f <- (10 - 1)*m.f.trunc + 1
lambda.F <- lambda*f.L/f.f
f.L ~ dbeta(1, 4)
f.f ~ dbeta(1, 2)
rho.L ~ dlnorm(mu.rho.L, tau.rho.L)
mu.rho.L <- log(mean.rho.L) - pow(sigma.rho.L, 2)/2
sigma.rho.L <- log(RF.rho.L)/1.645
tau.rho.L <- pow(sigma.rho.L, -2)
mu <- P.LD/(T.LI + T.R)
#Triangular distribution replaced with beta distribution over same range
P.LD.trunc ~ dbeta(9, 1)
P.LD <- (0.99-0.5)*P.LD.trunc + 0.5
omega <- P.I*P.FD/(T.FI + T.R)
```

(continued)

Table 13.16 (continued)

P.FD.trunc ~ dbeta(9, 1)

P.FD <- (0.99-0.1)*P.FD.trunc + 0.1

#Sum over appropriate rupture frequency range for given rupture size

#0.1-inch break

#rho.F <- sum(rho.rupt[2:8])/f.f

#10-inch break

rho.F <- sum(rho.rupt[6:8])/f.f

####################################

Markov system equations

solution[1:n.grid, 1:dim] <- ode(init[1:dim], times[1:n.grid], D(P[1:dim], t), origin, tol)

D(P[1], t) <- -phi*P[1] + omega*P[2] + mu*P[3]

D(P[2], t) <- phi*P[1] - lambda.F*P[2] - omega*P[2] - rho.F*P[2]

D(P[3], t) <- lambda.F*P[2] - mu*P[3] - rho.L*P[3]

D(P[4], t) <- rho.F*P[2] + rho.L*P[3]

#System hazard rate

#This is the rate at which ruptures occur divided by the probability of not being in the ruptured state at time j

for(j in 1:n.grid) {

        h.sys[j] <- (rho.F*solution[j,2] + rho.L*solution[j,3])/(solution[j,1] + solution[j,2] + solution[j,3])

        }

####################################

#Prior distribution parameters

mu.sc <- log(prior.mean.sc) - pow(sigma.sc, 2)/2

sigma.sc <- log(RF.sc)/1.645

tau.sc <- pow(sigma.sc, -2)

mu.dc <- log(prior.mean.dc) - pow(sigma.dc, 2)/2

sigma.dc <- log(RF.dc)/1.645

tau.dc <- pow(sigma.dc, -2)

}

data

list(

n.grid = 100, dim = 4, origin = 0, tol = 1.0E-8, init=c(1,0,0,0),

n.sc=c(8,8,8,8,8,8,8,8,8),

n.dc=c(4,4,4),

prior.mean.sc=4.27E-5, RF.sc=100,

prior.mean.dc=1.24E-6, RF.dc=100,

time.sc=c(423910, 84782, 16956, 282607, 56521, 11304, 141303, 28261, 5652),

w.sc=c(0.0625, 0.125, 0.0625, 0.125, 0.25, 0.125, 0.0625, 0.125, 0.0625),

time.dc=c(1695641, 1130428, 565213.8),

w.dc=c(0.25, 0.50, 0.25),

P.I=1, T.FI=10, T.R=200,

T.LI=1.5,

prior.mean.R=c(0.13, 0.06, 0.021, 5.0E-3, 8.0E-4, 1.8E-4, 3.8E-5, 2.11E-5),

RF.R=c(2.8, 3.4, 4.5, 6.6, 10.5, 15.1, 21.8, 25.1),

x.R=c(0,0,0,0,0,0,0,0), n.R=12,

mean.rho.L=0.02, RF.rho.L=3

)

Table 13.17 Posterior summaries of rate parameters in Markov model

	5th	50th	Mean	95th
λ_{SC}	1.7E-5	1.3E-4	2.7E-4	1.0E-3
λ_{DC}	8.7E-7	2.8E-6	3.5E-6	8.5E-6
$\lambda = \lambda_{SC} + \lambda_{DC}$	2.0E-5	1.3E-4	2.7E-4	1.0E-3
$\lambda_F = \lambda f_L / f_f$	2.9E-6	7.7E-5	1.6E-3	2.0E-3

Table 13.18 Posterior summaries of conditional rupture probabilities based on updating lognormal priors with 0 ruptures in 12 trials

Rupture size (in.)	5th	50th	Mean	95th
0.032	0.03	0.07	0.08	0.16
0.10	0.01	0.03	0.04	0.10
0.32	2.8E-3	0.01	0.02	0.05
1.00	3.8E-4	2.4E-3	4.3E-3	0.01
3.16	2.7E-5	2.8E-4	7.9E-4	2.9E-3
10.0	3.1E-6	4.7E-5	1.8E-4	6.9E-4
31.62	3.1E-7	6.7E-6	3.6E-5	1.4E-4
42.4	1.2E-7	3.1E-6	2.5E-5	8.0E-5

Table 13.19 Posterior summaries of rupture frequencies λ times conditional rupture probabilities from Table 13.18

Rupture size (in.)	5th	50th	Mean	95th
0.032	1.2E-6	9.4E-6	2.1E-5	8.4E-5
0.10	5.1E-7	4.6E-6	1.1E-5	4.5E-5
0.32	1.5E-7	1.6E-6	4.4E-6	1.8E-5
1.00	2.3E-8	3.3E-7	1.2E-6	4.8E-6
3.16	1.9E-9	3.9E-8	2.1E-7	8.1E-7
10.0	2.3E-10	6.3E-9	4.7E-8	1.8E-7
31.62	2.4E-11	9.0E-10	9.3E-9	3.5E-8
42.4	9.8E-12	4.2E-10	7.0E-9	2.0E-8

Table 13.20 Posterior summaries of remaining Markov parameters

	5th	50th	Mean	95th
$\phi = m_f \lambda$	5.4E-5	4.65E-4	1.1E-3	4.3E-3
μ	4.2E-3	4.7E-3	4.7E-3	4.9E-3
ω	3.5E-3	4.4E-3	4.3E-3	4.7E-3
ρ_L	5.3E-3	1.6E-2	2.0E-2	4.8E-2
ρ_F (0.1-in.)	2.6E-6	3.1E-5	8.1E-4	6.5E-4
ρ_F (10-in.)	2.2E-9	5.2E-8	2.0E-6	1.6E-6

Table 13.21 Results for system hazard rate (yr^{-1})

h(t, size)	5th	Median	Mean	95th
h(1, 0.1)	2.1E-10	1.5E-8	8.7E-7	1.6E-6
h(10, 0.1)	2.6E-9	1.9E-7	7.4E-6	1.9E-5
h(40, 0.1)	1.3E-8	1.0E-6	2.8E-5	9.9E-5
h(100, 0.1)	3.8E-8	2.8E-6	6.0E-5	2.5E-4
h(1, 10)	3.6E-12	3.6E-10	3.2E-8	4.9E-8
h(10, 10)	2.3E-10	2.7E-8	1.8E-6	3.9E-6
h(40, 10)	2.7E-9	3.2E-7	1.4E-5	4.3E-5
h(100, 10)	1.1E-8	1.3E-6	3.9E-5	1.5E-4

to the ODEs are shown in Tables 13.17, 13.18, 13.19, and 13.20. The hazard rate results for 10 inch ruptures are plotted in Fig. 13.7 and tabulated in Table 13.21.

References

1. Coles S (2001) An introduction to statistical modeling of extreme values. Springer, Berlin
2. Gumbel EJ (1958) Statistics of extremes. Columbia University Press, New York
3. R Development Core Team (2011) R: A language and environment for statistical computing. Vienna, Austria
4. Mosleh A (1992) Bayesian modeling of expert-to-expert variability and dependence in estimating rare event frequencies. Reliab Eng Syst Saf 38:47–57
5. Dezfuli H, Kelly DL, Smith C, Vedros K, Galyean W (2009) Bayesian inference for NASA probabilistic risk and reliability analysis. NASA, Washington
6. Siu NO, Kelly DL (1998) Bayesian parameter estimation in probabilistic risk assessment. Reliab Eng Syst Saf 62:89–116
7. Fleming KN (2004) Markov models for evaluating risk-informed in-service inspection strategies for nuclear power plant piping systems. Reliab Eng Syst Saf 83:27–45

Appendix A
Probability Distributions

A.1 Discrete Distributions

A.1.1 Binomial Distribution

The **binomial** distribution (Bernoulli model) describes the number of failures, X, in n independent Bernoulli trials. The random variable X has a binomial distribution if:

- The number of random trials is one or more and is known in advance.
- Each trial results in one of two outcomes, usually called success and failure (or could be pass-fail, hit-miss, defective-nondefective, etc.).
- The outcomes for each trial are statistically independent of the outcomes of other trials.
- The probability of failure, p, is constant across trials.

Equal to the number of failures (or successes depending upon the context) in the n trials, a binomial random variable X can take on any integer value from 0 to n, inclusive of the endpoints. The probability associated with each of these possible outcomes, x, is given by the binomial(n, p) probability density (mass) function (pdf) as

$$\Pr(X = x) = \binom{n}{x} p^x (1 - p)^{n-x}, \quad x = 0, \ldots, n$$

Here

$$\binom{n}{x} = \frac{n!}{x!(n - x)!}$$

is the binomial coefficient and $n! = n(n - 1)(n - 2) \ldots (2)(1)$, the factorial function, with $0!$ defined to be equal to 1. The binomial coefficient provides the number of ways in which exactly x failures can occur in n trials (number of

D. Kelly and C. Smith, *Bayesian Inference for Probabilistic Risk Assessment*,
Springer Series in Reliability Engineering, DOI: 10.1007/978-1-84996-187-5,
© Springer-Verlag London Limited 2011

Fig. A.1 Two binomial
probability density functions

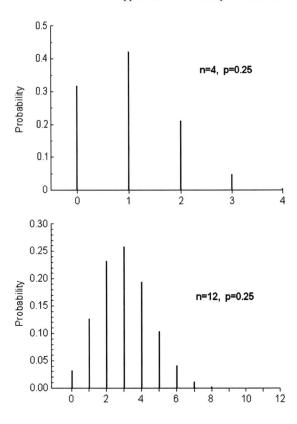

combinations of n trials selected x at a time). Examples of a binomial pdf are
shown in Fig. A.1.

The binomial distribution has two parameters, n and p, of which n is specified.
Note that, while n is specified, it can be subject to uncertainty in the Bayesian
framework.

The mean and variance of a binomial(n, p) random variable X are

$$E(X) = np$$
$$\text{Var}(X) = np(1 - p).$$

Figure A.1 shows two binomial pdfs, each with parameter $p = 0.25$, and $n = 4$
and 12. In each case, the mean is np. The means have been aligned in the two plots.

Software	Function
Excel	=BINOMDIST(x, n, p, cumulative)
	If cumulative = false then return pdf
	cumulative = true then return cdf
OpenBUGS	$\sim$dbin(p, n)

A.1.2 Poisson Distribution

The Poisson distribution describes the total number of events occurring in some interval of time(t) or space. The pdf of a Poisson random variable X, with parameter $\mu = \lambda t$, is

$$\Pr(X = x) = \frac{e^{-\mu}\mu^x}{x!} = \frac{e^{-\lambda t}(\lambda t)^x}{x!}$$

for $x = 0, 1, 2, \ldots$, and $x! = x\,(x - 1)\,(x - 2)\ldots(2)(1)$, as defined previously.

Several conditions are assumed to hold for a Poisson process that produces a Poisson random variable:

- For small intervals, the probability of exactly one occurrence is approximately proportional to the length of the interval (where λ, the event rate or intensity, is the constant of proportionality).
- For small intervals, the probability of more than one occurrence is essentially equal to zero (see below).
- The numbers of occurrences in two non-overlapping intervals are statistically independent.

The Poisson distribution has a single parameter μ, denoted Poisson(μ). If X denotes the number of events that occur during some time period of length t, then X is often assumed to have a Poisson distribution with parameter $\mu = \lambda t$. In this case, X is generated by a Poisson process with intensity $\lambda > 0$. The parameter λ is also referred to as the event rate (or failure rate when the events are failures). Note that λ has units 1/time; thus, $\lambda t = \mu$ is dimensionless.

The expected number of events occurring in the interval 0 to t is $\mu = \lambda t$. Thus, the mean of the Poisson distribution is equal to the parameter of the distribution, which is why μ is often used to represent the parameter. The variance of the Poisson distribution is also equal to the parameter of the distribution. Therefore, for a Poisson(μ) random variable X,

$$E(X) = \text{Var}(X) = \mu = \lambda t.$$

Figure A.2 shows two Poisson pdfs, with means $\mu = 1.0$ and 3.0, respectively. The means have been aligned in the plots. Note the similarity between the Poisson distribution and the binomial distribution when $\mu = np$ and n is fairly large.

The Poisson distribution is important because it describes the occurrence of many rare events, regardless of their underlying physical process. It also has many applications to describing the occurrences of system and component failures under steady-state conditions. These applications utilize the relationship between the Poisson and *exponential* distributions: the times between successive Poisson-distributed events follow an exponential distribution.

Fig. A.2 Two Poisson
probability density functions

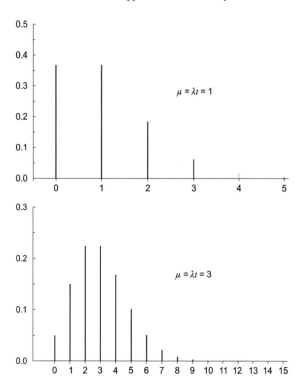

Software	Function
Excel	=POISSON(x, mu, cumulative)
	mu is the mean
	If cumulative = false then return pdf
	Cumulative = true then return cdf
OpenBUGS	$\sim$ dpois(lambda)
	Lambda is the mean

A.1.3 Multinomial Distribution

This is an extension to the binomial distribution, where there are more than two
possible outcomes. It is the aleatory model most commonly used for common-
cause failure. The observed data typically are in the form of a vector of event
counts, $n = (n_1, n_2,..., n_k)$, with $\Sigma n_i = N$. The pdf is given by

$$f(n_1, \ldots, n_m) = \binom{N}{n_1 n_2 \cdots n_m} \alpha_1^{n_1} \alpha_2^{n_2} \cdots \alpha_m^{n_m}$$

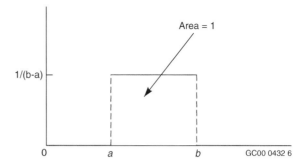

Fig. A.3 Uniform (a, b) distribution

where

$$\binom{N}{n_1 \cdots n_m} = \frac{N!}{n_1! \cdots n_m!}$$

The marginal means and variances are $E(n_i) = N\alpha_i$ and $Var(n_i) = N\alpha_i(1 - \alpha_i)$, respectively. The multinomial distribution is specified as a model in OpenBUGS as $n[1{:}m] \sim dmulti(alpha[1{:}m], N)$.

A.2 Continuous Random Variables

A.2.1 Uniform Distribution

A uniform distribution represents the case where any value in a specified interval $[a, b]$ is equally likely. For a uniform random variable, X, because the outcomes are equally likely, the pdf $f(x)$ is equal to a constant. The pdf of a uniform distribution with parameters a and b, denoted uniform(a, b) is

$$f(x) = \frac{1}{b - a}$$
$$\text{for } a \leq x \leq b.$$

Figure A.3 shows the density of the uniform(a, b) distribution.
The mean and variance of a uniform(a, b) distribution are

$$E(X) = \frac{b + a}{2}$$

and

$$Var(X) = \frac{(b - a)^2}{12}$$

Software	Function
Excel	$=(b - a) * \text{percentile} + a$
	Returns the value for the percentile specified
OpenBUGS	$\sim \text{dunif}(a, b)$

A.2.2 Normal Distribution

The normal distribution is characterized by two parameters, μ and σ. For a random variable, X, that is normally distributed with parameters μ and σ, the pdf of X is

$$f(x) = \frac{1}{\sigma\sqrt{2\pi}}\exp\left[-\frac{1}{2}\left(\frac{x - \mu}{\sigma}\right)^2\right]$$

for $-\infty < x < \infty$, $-\infty < \mu < \infty$, and $\sigma > 0$. Increasing μ moves the density curve to the right and increasing σ spreads the density curve out to the right and left while lowering the peak of the curve. The units of μ and σ are the same as for X.

The mean and variance of a normal distribution with parameters μ and σ are

$$E(X) = \mu$$

and

$$\text{Var}(X) = \sigma^2.$$

The normal distribution is denoted normal(μ, σ^2).

Software	Function
Excel	$=\text{NORMDIST}(x, \text{mean}, \text{sdev}, \text{cumulative})$
	If cumulative = false then return pdf
	Cumulative = true then return cdf
	$=\text{NORMINV}(\text{percentile}, \text{mean}, \text{sdev})$
	Returns the value for the percentile specified
OpenBUGS	$\sim \text{dnorm}(\text{mu}, \text{tau})$
	Note mu is the mean and tau = 1/variance

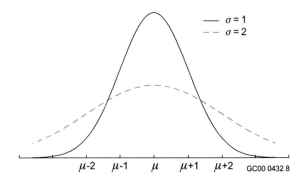

Fig. A.4 The normal distribution

A.2.3 Lognormal Distribution

The distribution of Y is lognormal when the distribution of $\ln(Y)$ is normal. The parameters of the lognormal distribution are μ and σ, the parameters from the associated normal distribution. For a variable, Y, that is lognormally distributed with parameters μ and σ, denoted lognormal(μ, σ^2), the pdf is

$$f(y) = \frac{1}{\sigma y \sqrt{2\pi}} \exp\left[-\frac{1}{2\sigma^2}(\ln(y) - \mu)^2\right]$$

for $0 < y < \infty$, $-\infty < \mu < \infty$, and $\sigma > 0$. The mean and variance of a lognormal (μ, σ^2) distribution are

$$E(Y) = \exp\left[\mu + \frac{\sigma^2}{2}\right]$$

and

$$Var(Y) = \exp\left(2\mu + \sigma^2\right)\left[\exp\left(\sigma^2\right) - 1\right].$$

The median of a lognormal distribution is $\exp(\mu)$ and the mode is $\exp(\mu-\sigma^2)$. A dispersion measure for the lognormal distribution that is commonly used in PRA is the error factor (EF), where

$$EF = \exp(1.645\sigma),$$

and is defined as

$$\Pr\left[\frac{med(Y)}{EF} \leq Y \leq med(Y) * EF\right] = 0.90.$$

To calculate probabilities for a lognormal(μ, σ^2) random variable, Y, the tables for the standard normal distribution can be used. Specifically, for any number b,

Fig. A.5 The lognormal distribution

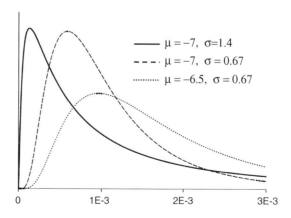

$$\mu = -7, \; \sigma = 1.4$$
$$\mu = -7, \; \sigma = 0.67$$
$$\mu = -6.5, \; \sigma = 0.67$$

0   1E-3   2E-3   3E-3

$$\Pr[Y \le b] = \Pr[\ln(Y) \le \ln(b)] = \Pr[X \le \ln(b)]$$
$$= \Phi\left[\frac{(\ln(b) - \mu)}{\sigma}\right]$$

where $X = \ln(Y)$ is normal(μ, σ^2).

Software	Function
Excel	=LOGNORMDIST(x, mean, sdev)
	Returns the cdf only
	=LOGINV(percentile, mean, sdev)
	Returns the value for the percentile specified
OpenBUGS	~ dlnorm(mu, tau)
	Note mu is the logarithmic mean and tau = 1/(logarithmic variance)

A.2.4 Logistic-normal Distribution

The logistic-normal distribution is not used widely in PRA, but it can be useful as a prior distribution for p in the binomial distribution, especially in settings where the values of p are expected to be large. Past PRAs have often used the lognormal distribution to represent uncertainty in p, and problems can be encountered in such cases, because values of the lognormal distribution are not constrained to be in the interval [0, 1]. The logistic-normal distribution can be an alternative to the lognormal distribution in cases where the heavy tail of the lognormal distribution (relative to a conjugate beta prior) is attractive, but which avoids the problem of having values of $p > 1$, because the values of the logistic-normal distribution lie in the interval [0, 1].

The pdf is given by

$$f(p) = \frac{1}{\sqrt{2\pi}\sigma p(1 - p)} \exp\left\{-\frac{\left[\ln\left(\frac{p}{1-p}\right) - \mu\right]^2}{2\sigma^2}\right\}$$

Like the lognormal distribution, μ and σ^2 are the parameters of an underlying normal distribution. In this case $\ln[p/(1 - p)]$ is normal(μ, σ^2). Just as percentiles of a lognormal distribution were related to percentiles of the underlying normal distribution, a similar relationship holds here. Percentiles of the logistic-normal distribution are given by

$$p_q = \frac{\exp(y_q)}{1 + \exp(y_q)}$$

where y_q is the $(q \times 100)$th percentile of the underlying normal distribution. For example, the 95th percentile of p is given by

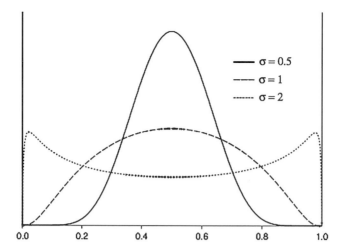

Fig. A.6 Logistic-normal densities, each with median equal to 0.5

$$p_{0.95} = \frac{\exp(\mu + 1.645\sigma)}{1 + \exp(\mu + 1.645\sigma)}$$

The moments of the logistic-normal distribution must be found numerically, as there is no closed-form analytic expression for the integrals involved. However, it is not difficult to find the moments with modern software.

The logistic-normal pdf can take on a variety of shapes, as shown in Fig. A.6. The logistic-normal distribution is not available directly in OpenBUGS, but can be entered by specifying a normal distribution for the underlying variable, and then defining the logistic-normal distribution in terms of the inverse logit transformation. For example, to enter a logistic-normal distribution with $\mu = -5$ and $\sigma = 1.2$, one would need the following lines of script in OpenBUGS:

```
p.norm ~ dnorm(mu, tau)
tau <- pow(sigma, -2)
p <- exp(p.norm)/(1 + exp(p.norm))
data
list(mu=-5, sigma=1.2)
```

A.2.5 Exponential Distribution

For a Poisson random variable X representing the number of events in a time interval t and for a random variable T defining the time between events, it can be shown that T has the exponential pdf

$$f(t) = \lambda e^{-\lambda t},$$

for $t > 0$. Thus, the time to the first event and the times between successive events follow an exponential distribution when the number of events in a fixed time interval follows a Poisson distribution.

The exponential distribution parameter, λ, is referred to as the failure rate if the events are component or system failures and the failures are repaired to a same-as-new condition after each failure. For the exponential distribution, the failure rate is constant. The cdf of the exponential distribution is

$$F(t) = 1 - e^{-\lambda t}.$$

The exponential distribution with parameter λ is denoted exponential(λ). The mean and variance of an exponential(λ) distribution are

$$E(T) = \frac{1}{\lambda}$$

and

$$\mathrm{Var(T)} = \frac{1}{\lambda^2}.$$

The relationship of the exponential distribution to the Poisson process can be seen by observing that the probability of no failures before time t can be viewed in two ways. First, the number of failures, X, can be counted. The probability that the count is equal to 0 is given as

$$\mathrm{Pr}(X = 0) = e^{-\lambda t} \frac{(\lambda t)^0}{0!} = e^{-\lambda t}$$

Alternatively, the probability that first failure time, T, is greater than t is

$$\mathrm{Pr}(T > t) = 1 - \mathrm{Pr}(T \le t) = 1 - F(t) = 1 - \left[1 - e^{-\lambda t} \right] = e^{-\lambda t}.$$

Thus, the two approaches give the same expression for the probability of no failures before time t.

The assumptions of a homogeneous Poisson process require a constant failure rate, λ, which can be interpreted to imply that the failure process has no memory. Thus, if a device is still functioning at time t, it remains as good as new and its remaining life has the same exponential(λ) distribution.

Software	Function
Excel	=EXPONDIST(x, lambda, cumulative)
	If cumulative = false then return pdf
	Cumulative = true then return cdf
OpenBUGS	~ dexp(lambda)

Fig. A.7 The exponential
distribution

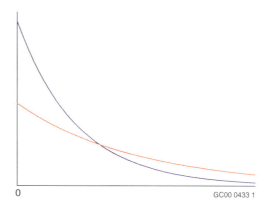

0   GC00 0433 1

A.2.6 Weibull Distribution

For a random variable, T, that has a Weibull distribution, the pdf is

$$f(t) = \frac{\beta}{\alpha} \left(\frac{t}{\alpha}\right)^{\beta-1} \exp\left[-\left(\frac{t}{\alpha}\right)^{\beta}\right],$$

for $t \geq 0$ and parameters $\alpha > 0$ and $\beta > 0$. The cdf for T is

$$F(t) = 1 - \exp\left[\left(\frac{t}{\alpha}\right)^{\beta}\right],$$

for $t \geq 0$.

OpenBUGS uses a different parameterization, obtained by defining $\lambda = \alpha^{-\beta}$. In this parameterization the pdf is given by

$$f(t) = \beta \lambda t^{\beta-1} \exp\left(-\lambda t^{\beta}\right)$$

and cdf given by

$$F(t) = 1 - \exp\left(-\lambda t^{\beta}\right)$$

The α parameter is a scale parameter that expands or contracts the distribution on the horizontal axis. The β parameter is a shape parameter and allows for a variety of distribution shapes. When $\beta = 1$, the Weibull distribution reduces to the exponential distribution, where $\lambda = 1/\alpha$. Therefore, the Weibull family of distributions includes the exponential family of distributions as a special case.

Fig. A.8 The Weibull
distribution

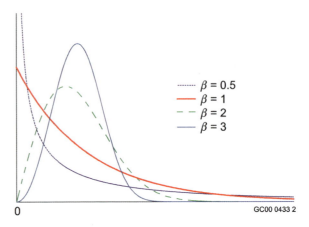

$\cdots\cdots$ $\beta = 0.5$
—— $\beta = 1$
– – $\beta = 2$
—— $\beta = 3$

0   GC00 0433 2

Software	Function
Excel	=WEIBULL(x, beta, alpha, cumulative)
	If cumulative = false then return pdf
	Cumulative = true then return cdf
	Note the ordering of beta and alpha in the function
OpenBUGS	~ dweib(beta, lambda)

A.2.7 Gamma Distribution

For a random variable, T, that has a gamma distribution, the pdf is

$$f(t) = \frac{\beta^{\alpha}}{\Gamma(\alpha)} t^{\alpha-1} \exp(-t\beta),$$

for t, α, and $\beta > 0$. Here

$$\Gamma(\alpha) = \int_0^{\infty} x^{\alpha-1} e^{-x} dx$$

is the gamma function evaluated at α. If α is a positive integer, $\Gamma(\alpha) = (\alpha - 1)!$

A gamma distribution with parameters α and β is referred to as gamma(α, β). The mean and variance of the gamma(α, β) random variable, T, are:

$$E(T) = \frac{\alpha}{\beta} \text{ and } Var(T) = \frac{\alpha}{\beta^2}.$$

The parameters α and β are referred to as the shape and scale parameters, respectively. The gamma($\alpha = n/2$, $\beta = \frac{1}{2}$) distribution is known as the chi-squared distribution with n degrees of freedom, denoted $\chi^2(n)$.

Fig. A.9 The gamma
distribution

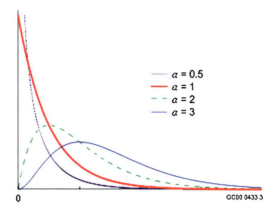

For the gamma distribution, when $\alpha < 1$, the density has a vertical asymptote at $t = 0$. When $\alpha = 1$, the gamma distribution reduces to an exponential distribution. When α is large, the distribution is approximately a normal distribution.

Software	Function
Excel	=GAMMADIST(x, alpha, 1/beta, cumulative)
	If cumulative = false then return pdf
	Cumulative = true then return cdf
	=GAMMAINV(percentile, alpha, 1/beta)
	Returns the value for the percentile specified
OpenBUGS	$\sim$ dgamma(alpha, beta)

A.2.8   Beta Distribution

The pdf of a beta random variable, Y, is

$$f(y) = \frac{\Gamma(\alpha + \beta)}{\Gamma(\alpha)\Gamma(\beta)} y^{\alpha-1}(1 - y)^{\beta-1},$$

for $0 \leq y \leq 1$, with the parameters α, $\beta > 0$, and is denoted beta(α, β). The gamma functions at the front of the pdf form a normalizing constant so that the density integrates to 1. The pdf can also be written as

$$f(y) = \frac{y^{\alpha-1}(1 - y)^{\beta-1}}{B(\alpha, \beta)}$$

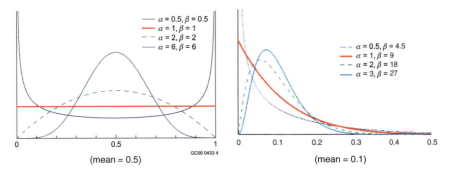

Fig. A.10 The beta distribution

where $B(\alpha, \beta)$ is the beta function, defined as

$$B(\alpha, \beta) = \int_0^1 x^{\alpha-1}(1-x)^{\beta-1}dx$$

The mean and variance of the beta(α, β) random variable, Y, are

$$E(Y) = \frac{\alpha}{\alpha + \beta}$$

and

$$Var(Y) = \frac{\alpha\beta}{(\alpha + \beta)^2(\alpha + \beta + 1)}$$
$$= \frac{mean(1 - mean)}{\alpha + \beta + 1}$$

When $\alpha < 1$, the density has a vertical asymptote at $y = 0$, and when $\beta < 1$, the density has a vertical asymptote at $y = 1$. When $\alpha = \beta = 1$, the density is uniform. When $\alpha = \beta = 0.5$, the density is U-shaped and is the Jeffreys noninformative prior for a binomial likelihood function.

Software	Function
Excel	=BETADIST(x, alpha, beta)
	Returns the cdf only
	=BETAINV(percentile, alpha, beta)
	Returns the value for the percentile specified
OpenBUGS	~ dbeta(alpha, beta)

A.2.9 Dirichlet Distribution

The Dirichlet distribution is the conjugate prior to the multinomial distribution, and is most commonly used in PRA as a prior distribution for the alpha factors in the alpha-factor model of common-cause failure. It is a multivariate extension of the beta distribution. Its pdf is given by

$$f(\alpha_1,\ldots,\alpha_m) = \frac{\Gamma(\alpha_1 + \cdots + \alpha_m)}{\Gamma(\alpha_1)\cdots\Gamma(\alpha_m)}\theta_1^{\alpha_1-1}\cdots\theta_m^{\alpha_m-1}, \quad \sum_{i=1}^{m}\alpha_i = 1$$

The marginal means are given by

$$E(\alpha_i) = \frac{\theta_i}{\theta_t}, \quad \theta_t = \sum_{j=1}^{m}\theta_j$$

The marginal variances are given by

$$Var(\alpha_i) = \frac{\theta_i(\theta_t - \theta_i)}{\theta_t^2(\theta_t + 1)}$$

The marginal distribution of each α_i is beta$(\theta_i,\ \theta_t - \theta_i)$. The Dirichlet distribution is specified in OpenBUGS as alpha[1: m] $\sim$ ddirich(theta[1: m]).

Appendix B
Guidance for Using OpenBUGS

B.1 WinBUGS and OpenBUGS

BUGS is the acronym for Bayesian inference Using Gibbs Sampling. WinBUGS is freely available software for implementing Markov chain Monte Carlo (MCMC) sampling. The open source version can be found at the URL: www.openbugs.info

Both versions are commonly referred to as WinBUGS or just BUGS.

OpenBUGS is commonly used independently but can be called from other programs through shell commands or from the open-source statistical program R through the R2WinBUGS or BRUGS packages for R. For more information on R, see the R Project homepage, www.r-project.org.

The OpenBUGS user manual (Spiegelhalter et al. 2007) comes with the program and is the basis for much of this appendix.

B.1.1 Distributions Supported in OpenBUGS

Over 20 distributions are supported in OpenBUGS. Discrete and continuous univariate and multivariate distributions are supported. Common distributions used in PRA that are supported include:

- Binomial: dbin(p, n)
- Poisson: dpois(mu)

  - Users will often have mu = λt

- Exponential: dexp(lambda)
- Weibull: dweib(v, lambda)
- Gamma: dgamma(r, mu)
- Uniform: dunif(a, b)

D. Kelly and C. Smith, *Bayesian Inference for Probabilistic Risk Assessment*, Springer Series in Reliability Engineering, DOI: 10.1007/978-1-84996-187-5, © Springer-Verlag London Limited 2011

- Beta: dbeta(a, b)
- Lognormal: dlnorm(mu, tau)

    - Tau = $1/\sigma^2$
    - $\sigma = \ln(EF)/1.645$

It is also possible to analyze user-defined distributions in OpenBUGS. See the OpenBUGS User Manual (Spiegelhalter et al. 2007) or Chap. 13 for information on how to do this.

B.1.2 OpenBUGS Script

OpenBUGS is a scripting language with a menu-driven interface (Thomas 2006). There are three parts to an OpenBUGS script: the model description, data section, and initial values. A sample script is shown below:

```
Script to update rate
model {                        #Model defined between {} symbols
  events ~ dpois(mu)           #Poisson distribution for number of events
  mu <- lambda*time            #Parameter of Poisson distribution
  lambda ~ dgamma(2.6,34)      #Prior distribution for lambda
}
Data                           #Observed data
list(events = 2,time = 14)
```

The model description includes the likelihood function, prior distribution, and any derived quantities (e.g., system reliability). The data can be written within the OpenBUGS script or input from a separate text file. The initial values can be written within the OpenBUGS script or input from a separate text file.

The # character is used for comments. Commenting scripts is highly encouraged.

B.1.3 Demonstration of OpenBUGS via an Example Analysis

Assume the frequency of an event (lambda) has a gamma(2.6, 34 yr) prior distribution. The likelihood function is a Poisson distribution with observed data consisting of 2 events in a 14 year period. The posterior distribution is gamma (2.6 + 2, 34 + 14 year) and the posterior mean of lambda is 4.6/48 year = 0.096 per year.

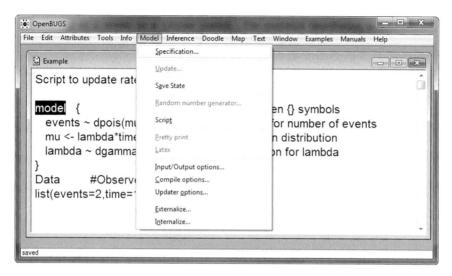

Fig. B.1 Checking the OpenBUGS script for syntax errors

The example script is used to update the prior distribution for lambda with the observed data. This can be written either in the OpenBUGS new project screen or cut and pasted from a text editor. The first step is to check the model's syntax by highlighting the word "model" and selecting "Model → Specification" from the toolbar as shown in Fig. B.1.

The Specification Tool will appear on the screen, after which the "Check Model" button should be selected. A status message will be displayed at the bottom left of the OpenBUGS palette. If no errors exist, the message is returned, "model is syntactically correct. "

Leave the Specification Tool on the screen and double-click (highlight) the word "list" in the data portion of the script, then select the "load data" button on the Specification Tool. A status message will be displayed at the bottom left of the OpenBUGS palette that states "data loaded." Next, select the "compile" button and OpenBUGS should report "model compiled." Note that models using multiple chains can be run. Only one chain is used in this example.

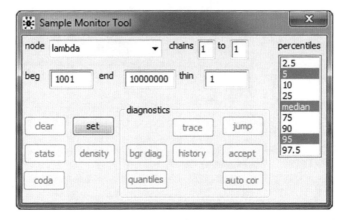

Fig. B.2 OpenBUGS Sample Monitor Tool

Fig. B.3 OpenBUGS update
tool

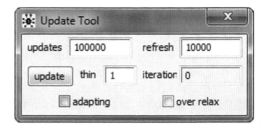

The last step with the Specification Tool is to load the initial values. For the conjugate prior example, the model is such a simple one that we can let OpenBUGS generate initial values. Select the "gen inits" button on the Specification Tool to have OpenBUGS generate the initial values and a message of "initial values generated, model initialized" should appear in the status bar.

The next step in setting up the analysis is to select the variables to monitor. For the conjugate prior example, we are interested in the frequency, or rate of occurrence (lambda). Close the Specification Tool and select "Inference → Samples" from the toolbar. The Sample Monitor Tool (Fig. B.2) will appear. Type the variable name "lambda" in the "node" box. Enter 1001 in the "beg" box as the iteration at which to start monitoring lambda (we are discarding the first 1,000 samples for burn-in). Save the setting by clicking the "set" button. Multiple variables and chains can be monitored, although we will only monitor lambda for this example.

Close the Sample Monitor Tool and select "Model → Update" from the toolbar. The Update Tool appears on the screen (Fig. B.3). Enter the number of updates to perform, in this case 100,000. The refresh rate will control the display in the iteration box during an update. A lower number for the refresh rate may slow down the sampling for problems that run especially quickly. Now select "update."

Fig. B.4 OpenBUGS
Sample Monitor Tool shown
the node lambda selection

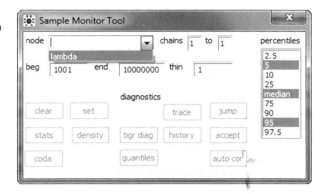

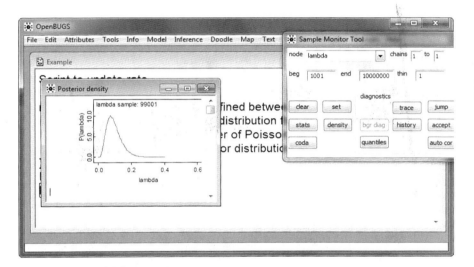

Fig. B.5 OpenBUGS density results for the node lambda

OpenBUGS will report that the "model is updating" in the status bar and the iteration window will display the iterations according to the increment specified in the refresh box. When the update has completed a message will appear in the status bar that indicates "updates took X s."

Close the Update Tool when the sampling is complete and re-open the Sample Monitor Tool by selecting "Inference → Samples" from the toolbar. Select "lambda" from the drop-down list of monitored nodes (Fig. B.4). Highlight the percentiles 5, median (50), and 95 by holding down the Ctrl key while selecting with the left mouse button. To display a graph of the posterior density (Fig. B.5) select "density".

Select the "stats" button to calculate the posterior mean and selected percentiles. To obtain compositional data (CoDa) for the posterior, select the "coda" button.

Note that the results (e.g., the density plots and the statistics) may be selected and copied (via the CTRL+C key combination in Windows) and pasted into other programs.

References

1. Spiegelhalter D, Thomas A, Best N, Lunn D (2007) OpenBUGS user manual, OpenBUGS project . http://www.mrc-bsu.cam.ac.uk/bugs
2. Thomas A (2006) The BUGS language, R News, vol 6/1, pp 17–21

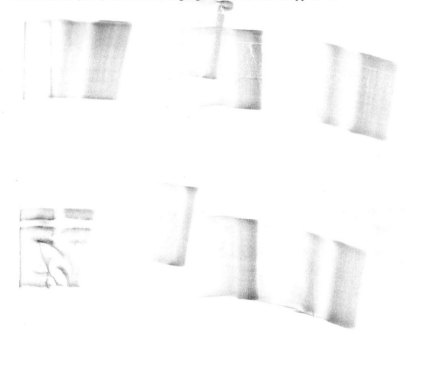

Index

D. Kelly and C. Smith, *Bayesian Inference for Probabilistic Risk Assessment*,
Springer Series in Reliability Engineering, DOI: 10.1007/978-1-84996-187-5,
© Springer-Verlag London Limited 2011